KING'S CROSS
TO ABERDEEN

Muskham Water Troughs 60908

V2 Class 2-6-2 engine no. 60908, seen here on the down line at Muskham Troughs between May and August 1958, was built at Darlington Works in April 1940 as part of an order for thirty-six locomotives intended for mixed traffic purposes. Upon nationalisation, the V2s received numbers starting at 60800. No. 60908 originally carried the LNER number 4879, which was changed to the BR number in November 1948. When new, the engine was equipped with a straight-sided high front tender which held 4,200 gallons of water. Fully loaded, the tender weighed approximately 54 tons. Doncaster Works actually produced more tenders than it needed for the 1939-40 order, so ten tenders were sent to Darlington and used on V2s produced there. No. 60908 was withdrawn in June 1962 and scrapped at Doncaster Works during the same month.

THE BILL REED COLLECTION

BRITISH RAILWAYS STEAM

KING'S CROSS TO ABERDEEN

PETER TUFFREY

FONTHILL

Grantham (South of the Station) 60977

South of Grantham, the track is level until it reaches Corby where the line rises at a gradient of 1:178. This continues until just before reaching Stoke tunnel where the line reaches its highest point between London and York at 345 feet above sea level. The line falls at a gradient of 1:200 in the 5-mile stretch from the tunnel and going through Great Ponton before reaching Grantham, which is situated 105 miles from London. The track south of Grantham is noteworthy for being the location where Class A4 locomotive no. 4468 *Mallard* broke the speed world record for steam locomotives in 1938. Gresley V2 Class 2-6-2 locomotive no. 60977, built at Darlington Works in October 1943, is seen on a fully fitted parcels train on the down line south of Grantham in May 1961. The engine left service in February 1962.

Fonthill Media Limited
Fonthill Media LLC
www.fonthillmedia.com
office@fonthillmedia.com

First published 2012

British Library Cataloguing in Publication Data:
A catalogue record for this book is available from the British Library

ISBN: 978-1-78155-053-3 (print)
ISBN: 978-1-78155-088-5 (e-book)

Typeset in 8.5 pt on 11 pt Sabon.
Printed and bound in England

Contents

Acknowledgements

I am grateful to the assistance received from the following people: Hugh Parkin, Bill Reed and Alan Sutton.

Special thanks are due to my son Tristram Tuffrey for his research and general help behind the scenes.

I have taken reasonable steps to verify the accuracy of the information in this book but it may contain errors or omissions. Any information that may be of assistance to rectify any problems will be gratefully received. Please contact me in writing: Peter Tuffrey, 8 Wrightson Avenue, Warmsworth, Doncaster, South Yorkshire, DN4 9QL.

Burnmouth 60861
Burnmouth is located just over the border into Scotland, 340 miles from London, 52 miles from Edinburgh and 72 miles from Newcastle. V2 Class 2-6-2 locomotive no. 60861 is seen leaving Burnmouth Station on 3 July 1961. The engine, modified to have separate cylinders, is in a neglected state with the later BR emblem.

Introduction

The photographs in this book were chosen from the many hundreds of black and white pictures taken by Bill Reed on and off the route stretching from King's Cross to Aberdeen. The majority of them were taken on 2¼ square negatives. Station scenes and views on works and in sheds are featured; they roughly cover a period from 1951 to 1967 and depict the last gasp of steam before the introduction of diesels. The book begins at King's Cross and finishes at Aberdeen as if on some imaginary journey.

Bill Reed was born in Nottingham in 1934 and during his younger days, once National Service was out of the way, he became a fireman on steam engines. Later, he became a driver on diesel locomotives. Throughout his life he has always taken black and white photographs, colour slides and also cine film of railways in the UK, Europe and America. A selection of the pictures he took around Nottingham formed the basis of an earlier book *The Last Days of Steam in Nottinghamshire from the Bill Reed Collection*, published by Alan Sutton in 2010. Another book entitled *Working and Preserved Industrial Locomotives from the Bill Reed Collection* appeared in 2012.

Many of the pictures reproduced here were taken by Bill after work at weekends or during annual holidays. Frequently using his rail passes, taking advantage of organised 'railway trips' or riding as a passenger on a friend's motorbike, he would find interesting locations where steam locomotives were still operating. It is amazing now to look at some of his pictures and realise he was standing on the railway track when taking them, unhindered by authority, yet always behaving responsibly. His proximity to his subjects can be breathtaking, as is amply demonstrated in this book.

This was clearly a time when trainspotting and taking pictures of steam locomotives was a revered pastime and not ridiculed as it is today. It was only natural for young lads to look at the vast, almost human engines with awe because maybe their dads, granddads or even great granddads had been part of building or operating them. For decades the railways had touched the lives of many families across the country.

It is noticeable that Bill has depicted marvellously the post Second World War atmosphere on the railways when steam was on its last legs; the vast majority of the locomotives are in a very grimy condition and a number are seen on the scrap line. There is also evidence of how complicated and labour intensive it was to run a steam engine; the vast coal hoppers and water tanks are just two examples.

How curious it must have been for loco drivers and firemen to move from the rigours of firing and raising steam on a steam loco one day and on the next step inside a cab of a diesel and press a button to make it work. The most remarkable exercise to my mind in the operation of a steam locomotive was replenishing its tender with water at troughs out on the track. How primitive this appears today; it is remarkably captured in operation by Bill at Muskham where water is cascading over the carriage nearest the tender.

Many of Bill's pictures were taken at the well known resting places for steam engines. For example, he was very active shooting film at Grantham, Doncaster, York, Darlington, Edinburgh Haymarket, Dundee and Aberdeen, and at times it was very difficult to know what to include and what to leave out of this book.

Not to be outdone when steam working was restricted in England during the early 1960s, Bill trekked with a friend, most notably Don Beecroft, to Scotland. In many instances they found places of interest where A4s and A1s were still operational, and a number are shown in the Edinburgh, Dundee and Aberdeen sections.

The new 'green' diesels shown here look remarkably modern and more efficient than their steam predecessors, and that is primarily why they have been included. But they too have given way to progress, being superseded by the electrics during the intervening years.

Knowing what he knows now about what happened to steam locomotives in later years, Bill says he would have taken many more pictures. But that is no matter; he has taken enough to give us more than a hint of what it was like in the last days of steam along the East Coast route.

CHAPTER ONE

King's Cross to Grantham

King's Cross 60009
Gresley A4 Class 4-6-2 locomotive no. 60009 *Union of South Africa*, photographed in June 1958 at King's Cross Station, was built at Doncaster Works in June 1937 and originally numbered 4488. The engine has the distinction of being the last A4 to leave on a scheduled service from King's Cross Station, departing for Newcastle on 24 October 1964. The locomotive was withdrawn in June 1966 and has been in the hands of various preservation societies since then.

Above: King's Cross 60505

Thompson Class A2/2 4-6-2 locomotive no. 60505 *Thane of Fife* is seen at King's Cross Station, Platform Two, in August 1958. Originally numbered 2005 as part of the P2 Class of locomotives, *Thane of Fife* was built at Doncaster in August 1936 with a 2-8-2 wheel configuration. Rebuilding of the locomotive, with a 4-6-2 wheel arrangement, occurred in January 1943; *Thane of Fife* was the first of its class to be rebuilt. It was also the first A2/2 to be withdrawn from service on 10 November 1959.

Opposite above: King's Cross 60014

Pictured at King's Cross shed on 15 October 1962, Class A4 Pacific 4-6-2 locomotive no. 60014 *Silver Link* was the first of the A4s to be completed by Doncaster Works, entering service on 7 September 1935. The engine's original number was 2509, changing to the British Rail number in June 1949. *Silver Link* was withdrawn from service on 29 December 1962 and was scrapped at Doncaster.

Opposite below: King's Cross 60067

Class A3 4-6-2 locomotive no. 60067 *Ladas* is at King's Cross shed on 15 October 1962. It was originally built as a Gresley Class A1 4-6-2 locomotive at NBLC Works, Glasgow, in August 1924. *Ladas* was rebuilt to A3 specifications in November 1939; the only such undertaking until 1941 due to the outbreak of the Second World War. The locomotive was withdrawn from service in December 1962.

King's Cross 60109

Another view of King's Cross shed on 15 October 1962 showing Class A3 4-6-2 locomotive no.
60109 *Hermit*, which was named after the 1867 Derby winner owned by H. Chaplin and bred by
William Blenkiron. The engine was built at Doncaster in July 1923, works no. 1570, and rebuilt to
A3 specifications in November 1943. *Hermit* was allocated to Doncaster and Gorton before moving
to King's Cross where it stayed until withdrawal from service in December 1962.

King's Cross 60110

After taking the 'top' photograph (on 15 October 1962), Bill Reed has moved to the right and
captured no. 60110 *Robert the Devil*. The engine was built at Doncaster in July 1923 and was
later given LNER Number 4479. It was rebuilt to A3 Class specifications during August 1942
and acquired a BR number in March 1949. Like *Hermit*, *Robert the Devil* is named after another
successful racehorse owned by C. Brewer, which won the St Leger at Doncaster in 1880. Withdrawal
from service came on 23 May 1963.

King's Cross 60134

The King's Cross shed building was originally constructed to serve as the Great Northern Railway Co.'s goods station for the London area. The 40 acres of land on which this site was created was purchased from St Bartholomew's Hospital for £40,000. Designs were prepared by Lewis Cubitt in 1849, with alterations being made to these by Edward Bury. The buildings were fully completed between 1851-52. Class A1 Peppercorn 4-6-2 locomotive no. 60134 *Foxhunter* is in steam at King's Cross shed on 15 October 1962. Built at Darlington in November 1948, the engine remained in service until October 1965.

King's Cross 60158

When this photograph was taken on 10 October 1962, King's Cross shed was host to approximately 160 locomotives. Peppercorn Class A1 4-6-2 locomotive no. 60158 *Aberdonian* was built at Doncaster Works in November 1949. The engine had two spells at King's Cross; the first when starting service in November 1949. *Aberdonian* also had allocations at Grantham and Copley Hill, Leeds, before returning to King's Cross in June 1957. It went on to Doncaster (36A as seen in the photograph) in September 1958 and remained there until withdrawal in December 1964.

King's Cross 60533

Peppercorn Class A2 4-6-2 locomotive no. 60533 *Happy Knight* is pictured at King's Cross, on 15 October 1962, in front of the coaler. This facility was constructed as part of an LNER improvements programme implemented between 1931 and late 1932, which aimed to increase the operational capabilities at King's Cross. With a 500-ton capacity, the coaler was located on the south side of the yard. The old coaling shed was subsequently demolished and replaced by offices. The 'new' coaler was demolished around 1964. *Happy Knight* was constructed at Doncaster in April 1948 and withdrawn from service in June 1963.

Opposite above: Peterborough New England Shed 61891

K3/2 Class 2-6-0 locomotive no. 61891 emerged from Darlington Works in July 1930. Designed by Nigel Gresley, the initial K3 locomotives were classified H4 by the GNR. The first ten were produced at Doncaster during 1920 and quickly proved to be successful at both goods work and passenger traffic. As a result, further lots of locomotives were produced by various companies; the last, no. 61992, was erected at Darlington in February 1937. A total of 193 engines were built, with no. 61891 being in lot 4. The engine was withdrawn from service in September 1961.

Below: **Peterborough New England 60513**

All but one of the fifteen locomotives in the Thompson A2/3 Class were named after racehorses that had triumphed at major races during the early 1940s. The exception was no. 60500 which took the name of retiring locomotive designer Edward Thompson. Class A2/3 4-6-2 locomotive no. 60513 *Dante*, built at Doncaster in August 1948, was named after the 1945 Derby winner owned by Sir Erik Ohlson. *Dante* spent time at King's Cross shed, New England and Grantham before final allocation to New England in June 1959. *Dante* left service in April 1963.

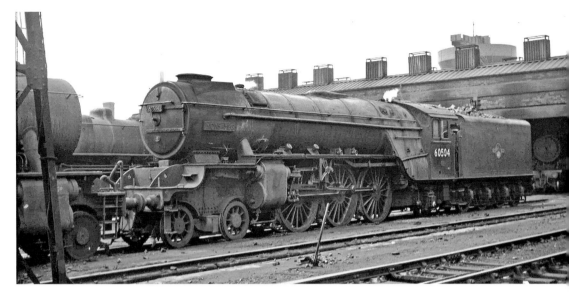

Peterborough New England 60504

Mons Meg was built at Doncaster to Gresley's P2 Class specifications in July 1936, and then rebuilt to Thompson's A2/2 criteria in November 1944 – again at Doncaster. During their time in service, the P2s encountered operating problems such as breaks to connecting rod bearings, crank axles and heating to the axleboxes. This led to the rebuilding of all six of the class between 1943-44. *Mons Meg*, seen here at New England shed, was withdrawn in January 1961.

Peterborough New England 60021

Construction of the A4s began in September 1935 to the specifications of Sir Nigel Gresley. All of the A4s were erected in Doncaster; the last emerging in July 1938. The cost of construction was in the region of £8,000 to £9,000. Class A4 locomotive no. 60021 *Wild Swan* is pictured with a late BR crest at Peterborough's New England shed on 8 December 1963. The engine was built at Doncaster in February 1938 with works no. 1869. *Wild Swan* was withdrawn in October 1963 and cut up at Doncaster Works during January 1964.

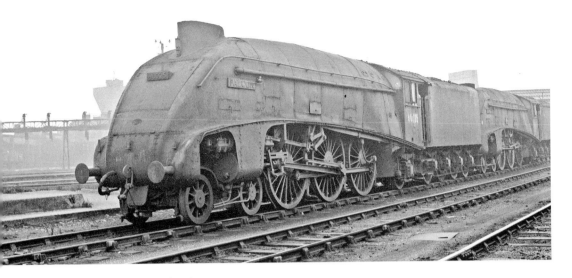

Peterborough New England 60025

Gresley Class A4 Pacific 4-6-2 locomotive no. 60025 *Falcon* emerged from Doncaster Works in February 1937. Thereafter, the locomotive was on shed at Edinburgh Haymarket before starting a series of moves between King's Cross and Grantham from March 1939 to May 1950. *Falcon* spent the next thirteen years at King's Cross until finally moving to Peterborough New England in June 1963. The locomotive was withdrawn four months later and cut up at Doncaster in January 1964. The photograph was taken on 8 December 1963.

Peterborough New England 60029

Woodcock was the name originally assigned to Class A4 Pacific 4-6-2 locomotive LNER no. 4489, later BR 60010, constructed in May 1937. However, the name was only in use on this locomotive for approximately a month before it was changed in June. It subsequently became *Dominion of Canada* allowing the name *Woodcock* to be assigned to LNER no. 4493, built in July 1937, and later numbered 60029. *Woodcock* was withdrawn from service in October 1963 and was scrapped at Doncaster Works by January 1964. The photograph was taken on 8 December 1963.

Peterborough New England 60063

New England shed came into service in 1852 and could accommodate up to twenty-four engines. Three years later, two extra lines were added to the existing six to increase capacity. Further extensions took place in 1866 to provide the shed with nine lines, which is how it remained until closure in 1968. Gresley A1 Class 4-6-2 locomotive no. 60063 *Isinglass* was built at Doncaster during July 1925 as LNER no. 2562. It is pictured on 8 December 1963 on the east side of the New England shed, with the fitting and wagon repair shops to the right. The locomotive was withdrawn in June 1964.

Opposite above: **Peterborough New England 60106**

The A3 Class was the result of a series of modifications made to the designs of the A1s which included increased boiler pressure, smaller cylinders and improved distribution of weight. Rebuilding occurred on all the original A1s and this process lasted until 1949. Built at Doncaster in April 1923, Gresley A1 Pacific Class 4-6-2 locomotive no. 60106 *Flying Fox* was rebuilt as an A3 in March 1947. One of the last of its class to be withdrawn at the end of December 1964, the engine was dismantled in Norwich in February 1965. Bill Reed took the picture on 8 December 1963.

Below: Peterborough New England 60140

The Peppercorn A1s entered service in August 1948, but without going through the tradition of being named (only one was named upon entering service – 60114 *W.P. Allen*). However, by 1950 this decision had been reversed and the naming of the class began. It followed naming practices such as adopting racehorse names, carrying on names from other locomotives and birds. A1 Peppercorn Class 4-6-2 locomotive no. 60140 *Balmoral*, entered traffic from Darlington in December 1948 with works no. 2059. It was given the name of *Balmoral* in July 1950, joining a number named after places in Scotland. Photographed on 8 December 1963, *Balmoral* was withdrawn in January 1965.

Peterborough New England D6806

This Class 37 diesel locomotive is pictured at the old coal stage at Peterborough New England shed with the mechanical coal stage just visible above the coal stage roof. The locomotive was built at English Electric Company's Vulcan Foundry, Newton-le-Willows and entered service in mid-January 1963. Initially, D6806 was allocated to Sheffield Darnall shed, which is where the locomotive was allocated when the photograph was taken in December 1963. D6806 was one of 309 Class 37s built for British Railways from 1960 to 1965 for use on both passenger services and freight duties. D6806 is seen with spilt headlamp code box, which is a feature that is not the same throughout the class. Differences include the headlamp code box placed centrally and on some not at all, with the box replaced by headlights. At the TOPS renumbering, D6806 became 37 106, receiving the number in February 1974. Some of the class are still in operation at present, while some have been preserved and the remainder have been scrapped. D6806, as 37 106, left service from Toton Traction Maintenance Depot, Nottinghamshire, in March 1999, after just over thirty-six years in service. It was subsequently scrapped at English, Welsh & Scottish Railways' Wigan scrapyard in August 2000.

Opposite above: Peterborough 63953

Doncaster Works built Gresley's 02/3 Class 2-8-0 locomotive no. 63953 in July 1932, works no. 1779. When entering service, the locomotive originally carried LNER no. 2960. This changed to no. 3953 in March 1946, before acquiring BR no. 63953 in June 1949. Seen here at Peterborough on 30 April 1955, the locomotive left service in November 1960, being scrapped at Doncaster the following month.

Below: **Peterborough 67392**

Peterborough's association with the railways began in the late 1840s with four railway companies, operating services from the town, sharing the station. The GNR opened its own dedicated railway station in August 1850 with a permanent engine shed replacing a temporary shed. The permanent Peterborough shed was constructed just after the opening of the station and had space for sixteen locomotives. Displaying an early BR crest at Peterborough's New England shed on 30 April 1955, C12 Class 4-4-2 locomotive no. 67392 was built at Doncaster Works in July 1907. The engine was withdrawn from service at the end of October 1956 and scrapped shortly after.

Peterborough 92039
Operations at Peterborough shed were very brief after the development of the New England shed in 1852. Despite this, the older shed continued to play a part, housing some passenger engines. However, with further expansion at New England shed during the mid-1900s, all responsibility for housing engines was transferred from Peterborough shed to New England. Photographed on 30 April 1955, Class 9F 2-10-0 locomotive no. 92039 was built at Crewe Works during December 1954 and removed from service in October 1965.

Peterborough 92043
The building of 251 Class 9F engines from 1954 to 1960 was shared between Crewe Works (198) and Swindon Works (58). Designed by Robert Riddles, the engines were intended to pull heavy freight trains over long distances but also found themselves on passenger duties during their time in service. Class 9F 2-10-0 locomotive no. 92043, built at Crewe Works in January 1955, was withdrawn in July 1966. The last working member of the class was withdrawn by mid-1968 though nine have survived into preservation. The photograph was taken on 30 April 1955.

Peterborough New England 60862

Doncaster Works erected the first five of the V2 Class between June and November 1936, and it eventually totalled some 184 locomotives. Pictured at Peterborough's New England shed, V2 Class 2-6-2 locomotive no. 60862 was built at Darlington Works in June 1939. It was one of 159 V2s built there between 1937 and 1944. No. 60862 was in service for twenty-five years before being withdrawn in June 1963.

Peterborough New England Shed 64223

Class J6 0-6-0 locomotive no. 64223 is pictured in front of the old coal stage at Peterborough New England shed. The facility was built as part of a series of improvements made during the early part of the twentieth century and replaced the coke stage from 1852 and a temporary coal stage erected in 1901. By 1933, the coal stage was replaced by a mechanical coaler similar to the one found at King's Cross shed. The coal stage remained intact until clearing of the New England site occurred in 1969. No. 64223 was built at Doncaster Works in October 1913 and withdrawn in April 1961. It was cut up at Doncaster during the following month.

Grantham 60134

Peppercorn A1 Class 4-6-2 locomotive no. 60134 *Foxhunter* is pictured at one of Grantham's ash pits located next to the coaling stage, which was at the south-west side of the old engine shed. The coaling stage was constructed during 1897 as part of a series of improvements at Grantham, which cost the GNR approximately £28,000. This included the setting of the foundations, the new engine shed for twenty engines (which could be increased by another twenty), the coaling stage and engine hoist. In the background, the imposing frame of the mechanical coal plant can be seen. This was put in place in the very early part of 1937 and was constructed by Henry Lees & Co. Ltd. for approximately £5,600. Built at Darlington Works in November 1948, *Foxhunter* was withdrawn in October 1965 and scrapped by the end of November.

Grantham 67773

Thompson L1 Class 2-6-4 locomotive no. 67773 is pictured at Grantham Station. The engine was built by Robert Stephenson & Hawthorns in December 1949 as part of an order for thirty-five engines placed in July 1947 by the LNER. Construction started in September 1949. Other builders included NBLC, thirty-five; Darlington Works, twenty-nine; and Doncaster Works, one, making a total of one hundred in the class. Thirty-eight of the one hundred engines produced had Westinghouse brakes, with the remaining sixty-two, including no. 67773, being equipped with steam brakes. No. 67773 was withdrawn in December 1962.

Grantham 43090

Darlington Works built Class 4MT 2-6-0 locomotive no. 43090 in December 1950. The design was originally made for the LMS by H. G. Ivatt as Class 4F, of which three were built for the company before it was absorbed into BR. One hundred and fifty-nine were built after nationalisation at Horwich Works, Darlington Works and Doncaster Works. This class was designed for mixed traffic work such as freight and local passenger services. No. 43090 is seen, on 13 April 1962, with a single chimney, which was fitted at the time of construction on this, and 109 other locomotives in its class. The first fifty were fitted with double chimneys but these hindered the effective operation of the locomotive and were removed and replaced with single chimneys. No. 43090 was withdrawn in April 1965.

Grantham 60033

Gresley's Class A4 4-6-2 Pacific no. 60033 *Seagull*, pictured at Grantham Station in July 1956, was built at Doncaster Works in June 1938. The engine is seen operating *The Northumbrian* service from London King's Cross to Newcastle Central and which ran from 1949 to 1963. The shed code 34A can be seen on the smokebox, which indicates that King's Cross was *Seagull's* allocation at the time. The locomotive was housed at King's Cross from construction until it was transfered to Grantham in April 1944. It returned south again in March 1948. *Seagull* was withdrawn at the end of December 1962 and scrapped at Doncaster Works in January 1963. The view shows the bay platform on the up side to the left, which has since been removed.

Grantham 60017

Gresley's Class A4 4-6-2 locomotive no. 60017 *Silver Fox*, pausing at Grantham Station in July 1956, was erected at Doncaster Works in December 1935 as LNER no. 2512. The first four A4s produced carried 'silver' in their names, as *Silver Fox* does. This is due to the locomotives being put to work on the *Silver Jubilee* service when they were built. These four locomotives also had liveries consisting of three shades of grey. No. 60017 received the green livery that it's pictured in during December 1951, having carried blue and black liveries since its original grey. The engine also carries a stainless steel fox donated by steelmaker Samuel Fox & Co. on its side and is visible here, as is the early BR emblem on the tender. *Silver Fox* was withdrawn in October 1963 and was scrapped at Doncaster by early December.

Opposite above: **Grantham 60047**

Pictured at Grantham on 21 January 1961, Gresley's 4-6-2 Pacific locomotive no. 60047 *Donovan* was built at Doncaster Works as an A1 in August 1924. The engine was rebuilt to A3 Class specifications during December 1947. *Donovan* is seen with a banjo dome above the boiler, indicating the boiler type fitted was diagram 94A. *Donovan*'s boiler was changed to a diagram 94HP during December 1957, but reverted back to a 94A during July 1959. Here, *Donovan* has a double chimney, which replaced the originally fitted single chimney during July 1959. The locomotive was withdrawn from service in April 1963.

Below: **Grantham 60051**

Easing into Grantham Station in July 1956, Gresley's 4-6-2 locomotive no. 60051 *Blink Bonny* was built as an A1 at Doncaster Works in October 1924 and rebuilt to A3 in November 1945. The engine is carrying the shed code 37B, which denotes Leeds Copley Hill shed and where it spent three years between May 1954 and September 1957 before moving to Heaton shed. Interestingly, no. 60051 had a varied allocation over its working life, spending time at Grantham, Peterborough New England, Gorton, King's Cross, Neasden, Gateshead and Darlington. The locomotive was allocated to Grantham for the longest period – eighteen years between October 1924 and January 1942. Withdrawal came in November 1964.

Above: Grantham 60105

Gresley's 4-6-2 locomotive no. 60105 *Victor Wild* is seen at Grantham with smoke deflectors, fitted in December 1960, and a double chimney, fitted in March 1959. The locomotive was converted to left-hand drive during February 1953. Originally built at Doncaster as an A1, in March 1923, *Victor Wild* was rebuilt to A3 specifications in October 1942. The locomotive was withdrawn in June 1963 and scrapped at Doncaster Works by the end of August.

Grantham 60867

The Ambergate Company was the first railway company to operate a service from Grantham, which started in July 1850 and travelled a route to Nottingham. The GNR main line service did not start operating until the end of the summer in 1852 when the section of track between Peterborough and Retford was finally completed. The Ambergate Company's station was made obsolete because of this development and all rail services started to operate from the GNR station. Gresley's V2 Class 2-6-2 locomotive no. 60867, built at Darlington in July 1939, is seen arriving at Grantham Station with a down express on 20 August 1958. It is equipped with a group standard straight-sided-type tender and was withdrawn in May 1962.

Opposite below: **Grantham 60125**

Built at Doncaster in April 1949, Peppercorn A1 Class 4-6-2 Pacific locomotive no. 60125 *Scottish Union* is pictured on 5 July 1957 at the ash pit at Grantham. This facility was located adjacent to the storage road on the left. In the background, the mechanical coaling stage is also visible. No. 60125 is carrying the early BR emblem on the tender while on the smokebox Grantham's shed code 35B can be seen. The engine had two spells at Grantham, the first being four months between February and June 1953 after which it moved to Leeds Copley Hill. *Scottish Union* returned in May 1954 and stayed for just over three years until being allocated to King's Cross in June 1957. Withdrawal came in July 1964.

Grantham 6157

Class B12 4-6-0 locomotive no. 61574 was built by Beyer, Peacock & Co. in August 1928 as part of the last ten locomotives manufactured for the class. The last ten differed in design from the previous seventy as they had no decorative trim over the wheels, Lentz valve gear and extended smokeboxes. They were classed B12/2. However, problems arose with the Lentz valve gear and it was removed from all locomotives (no. 61574's during January 1932). The engine was rebuilt to B12/3 in July 1933, receiving a larger boiler and other modifications. Photographed on 2 May 1955 the engine was withdrawn in January 1957.

Grantham 62000

Locomotive no. 62000 was originally built at Doncaster Works in June 1897 as an Ivatt GNR 4-4-0 Class D2. It was rebuilt in August 1916 with a Gresley boiler becoming Class D3. No. 62000 had a varied allocation, operating from Ardsley, Copley Hill, Darlington, Barnard Castle (1933-35), Hull (October 1935 to 1937) and then in the southern area. During repairs in 1944, the locomotive was assigned to operate officers' special trains; it did this from Grantham where it remained until it was the last of the class to be withdrawn in October 1951. The locomotive is pictured earlier in 1951 next to the coal stage and is fitted with windows in the cab.

Grantham 63939

Class 02/2, 2-8-0 locomotive, no. 63939, built at Doncaster Works in December 1923, is pictured during May 1960 at the engine hoist at Grantham shed. The facility was located next to the new engine shed at Grantham and was installed in 1897. A total of sixty-seven Class 02 locomotives were built between 1918 and 1943 for freight purposes. The Class was split into four sub-classes, with one locomotive classed as 02, ten as 02/1, fifteen as 02/2 and forty-one as 02/3. The 02/4 sub-class was formed when thirty of the engines were rebuilt. The 02/2 portion of the class had the cabs altered to fit the loading gauge at the time of Grouping. No. 63939 was removed from service in September 1963.

Grantham D1514

British Rail Class 47 diesel-electric locomotive D1514 was built by Brush Traction at their Falcon Works at Loughborough, and entered service during March 1963. The locomotive was built as part of the first batch of twenty engines ordered by BR; the class would eventually total 512 by 1968. D1514 was allocated to Finsbury Park from entering service before moving to Tinsley, York, Finsbury Park, Gateshead and back to Finsbury Park. The engine was withdrawn from Gateshead in April 1987 and scrapped in January 1990. The coaling plant and coal stage seen at the rear were demolished in November 1964 and during 1965 respectively.

Grantham 67362

C12 Class 4-4-2T locomotive no. 67362 was originally built at Doncaster Works during March 1899 as an Ivatt GNR C2 Class. The locomotives had two cylinders, Stephenson valve gear, and a diagram 11 boiler at 170 lbs of pressure. The engines weighed approximately 62 tons fully laden, with a capacity for 2.5 tons of coal and 1,350 gallons of water. No. 67362 has a short chimney, curved tank corners and an early BR emblem; it is pictured at Grantham Station on 5 July 1957 in front of the old engine shed. Withdrawal came in January 1958.

Grantham (South of the Station) 60500

Thompson's Class A2/3 Pacific 4-6-2 locomotive no. 60500 *Edward Thompson* is seen on a passenger service on the down line south of Grantham Station. The locomotive was built at Doncaster during May 1946 and had the distinction of being the 2,000th locomotive to be produced at the works. A total of fifteen A2/3 engines were built to Edward Thompson's specifications before he retired in June 1946, with thirty being planned originally. The last fifteen were redesigned by Arthur Peppercorn when he became Chief Mechanical Engineer of the LNER and classified Peppercorn A2. Photographed during May 1961, no. 60500 was removed from service in June 1963.

Grantham D208

The Vulcan Foundry at Newton-le-Willows built Class 40 diesel-electric locomotive D208 in early 1958. The class was first known as English Electric Type 4s and D208 was part of ten prototypes ordered by British Railways to test their suitability to different lines. The Great Eastern and East Coast Main Lines were chosen to test the engines but neither proved particularly suitable to the Type 4s. The West Coast Line was an exception and many found home there. Photographed on 12 September 1958, D208 was withdrawn in November 1982 and scrapped at Crewe Works in June 1988.

Grantham D356

Class 40 diesel-electric locomotive D356, built at Vulcan Foundry by English Electric was one of 170 produced by the company between 1958 and 1962. Twenty were also produced at Robert Stephenson & Hawthorns, bringing the total in the class to 200. D356 is seen outside Grantham on 14 April 1962 with a four-letter headcode box, which was fitted to the last fifty-nine locomotives produced. The headcode indicates that it is a freight train limited to 45 mph (7) destined for the North East (N) and 76 denoting the service number. D356 was removed from service in July 1980 and scrapped at Swindon Works by the end of the year.

Grantham D9003

BR Class 55 Deltic locomotive D9003 *Meld* was built by English Electric at the Vulcan Foundry between 1961 and 1962 as part of a batch of twenty-two ordered by BR. Entering service on 27 March 1961, D9003 is pictured at Grantham Station on 22 September 1962. The Class 55 locomotives were specifically used on the East Coast Main Line for passenger services. However, they had a relatively short lifespan as the arrival of high-speed trains on the line in the late 1970s and early 1980s meant some were in operation for less than twenty years. Sixteen were scrapped, including D9003, which was withdrawn at the end of 1980, and six survived into preservation.

Grantham (South of the Station) 60049

Gresley Class A1 4-6-2 Pacific (later A3 Class) locomotive no. 60049 *Galtee More* is pictured south of Grantham Station on the down line operating a passenger service in May 1961. The engine's smoke deflectors were fitted in October 1960 to improve visibility problems caused by the double chimney, which was added in March 1959. *Galtee More* is also carrying the GNR-type tender with coal rails with the later BR emblem.

CHAPTER TWO

Muskham Water Troughs to York

Muskham Water Troughs 60007

A4 Pacific 4-6-2 locomotive no. 60007 emerged from Doncaster Works in December 1937 as LNER no. 4498. It was the hundredth to be constructed to the Pacific design and to commemorate this it was suggested that the locomotive should be named after the designer Sir Nigel Gresley. No. 60007 is pictured on the up line at Muskham Troughs in August 1958. The troughs at Muskham were located just past Newark and approximately 120 miles from London. No. 60007 was withdrawn from service during February 1966 and went into the hands of the A4 Preservation Society, now the Sir Nigel Gresley Locomotive Preservation Trust Ltd.

Muskham Water Troughs 60042

Doncaster Works built Gresley Class A3 4-6-2 locomotive 60042 *Singapore* in December 1934. The engine is pictured on the down line with a north-bound express picking up water at Muskham Troughs just north of Newark. *Singapore* was originally fitted with a new-type non-corridor tender, which did not have coal rails but extended sides that curved inwards at the top. The GNR-type and new-type non-corridor tenders shared the same capacity of 5,000 gallons of water and 8 tons of coal. No. 60042 also carried a streamlined tender between February and November 1937. The final change to a GNR-type tender came in August 1943. The locomotive was withdrawn in July 1964.

Muskham Water Troughs 60118

There were six water troughs along the East Coast Main Line with Muskham being the third on the route travelling north. These troughs were 41 miles from Werrington Troughs to the south and 24 miles from Scrooby Water Troughs to the north. Peppercorn Class A1 4-6-2 locomotive no. 60118 *Archibald Sturrock* was a Doncaster-built A1, emerging from there in November 1948. It is pictured on the down line at Muskham during August 1958. The engine was named in July 1950 after mechanical engineer Archibald Sturrock who was the GNR locomotive superintendent during the 1850s. No. 60118 was withdrawn in October 1965 and scrapped by December.

Muskham Water Troughs 60119

Peppercorn Class A1 4-6-2 locomotive no. 60119 *Patrick Stirling* also takes its name from a GNR superintendent and acquired the name around the same time as no. 60118. The locomotive was built at Doncaster during November 1948 with works no. 2036. *Patrick Stirling* is pictured on the up line at Muskham Troughs in August 1958 with the later British Rail emblem on the tender, which was used between 1956 and 1967. The tenders of the Doncaster-built Peppercorn A1s are notably different from the ones built at Darlington as the Doncaster tenders had rivets protruding from the body. Those built at Darlington had tenders with rivets level with the body. No. 60119 was withdrawn in May 1964 but was not scrapped until August 1965.

Muskham Water Troughs 60143

Peppercorn Class A1 4-6-2 locomotive no. 60143 *Sir Walter Scott* was a Darlington-built A1 and emerged from there in February 1949. The 52A shed plate on the front of the smokebox indicates the locomotive was operating from Gateshead shed. *Sir Walter Scott* was allocated to Gateshead from new until moving to Heaton in May 1960. From there, the locomotive transferred to Tweedmouth and then to York before withdrawal in May 1964. The photograph dates from May 1958 and shows the locomotive on the down line with a north-bound express.

Above: **Muskham Water Troughs 60146**
Peppercorn Class A1 4-6-2 locomotive no. 60146 *Peregrine* entered traffic from Darlington during
April 1949 with works no. 2065. The locomotive is seen on the up line at Muskham Troughs with
a south-bound express in August 1958. Withdrawal of the Peppercorn A1s started in October 1962
with no. 60123 *H.A. Ivatt*, and by mid-1965 only thirteen of the forty-nine built were still in service.
No. 60146 was part of ten Peppercorn A1 withdrawals during October 1965, with a further one
being withdrawn in November. The remaining two lasted until March 1966 with no. 60145 *Saint
Mungo* being reprieved in April and lasting until June 1966.

Muskham Water Troughs 61331

At the time of Grouping in 1923, water troughs on the East Coast Main Line did not have a standard setting and varied considerably. Due to uncertainty over the setting of the tender scoop, levels of the troughs and the distance the scoops were to drop had to be standardised. During mid-1924, LNER mechanical engineers agreed that the top of the troughs should extend two inches above rail level and the tender scoop should drop no more than one inch below rail level. The changes were implemented by the end of the year with Muskham being the third trough on the line to be changed. Thompson B1 Class 4-6-0 locomotive no. 61331 is pictured on the up line at Muskham Troughs with a fully fitted freight train in August 1958. The engine was built by the NBLC, Queens Park Works, in June 1948. No. 61331 was withdrawn in September 1963.

Opposite below: Muskham Water Troughs 60810

The Muskham Troughs were the longest along the East Coast Main Line at 2,116 ft. The next was Scrooby with a length of 2,112 ft, followed by Wiske and Lucker, both at 1,869 ft. The shortest on the line were at both Langley and Werrington, measuring 1,780 ft in length. V2 Class 2-6-2 locomotive no. 60810, built at Darlington Works in September 1937, is collecting water on the down line at Muskham during August 1958. In service for twenty-eight years before withdrawal in November 1965, the engine was scrapped by January 1966.

Above: **Muskham Water Troughs 60966**

V2 Class 2-6-2 locomotive no. 60966 entered traffic from Darlington during March 1943. Doncaster Works was originally given the task of manufacturing the final two orders of V2 locomotives. Ten were ordered in early 1940 with a further twenty-five in April the following year. However, the LNER cancelled the Doncaster orders and switched them to Darlington Works. These orders were issued in August 1941 and building commenced on the final V2s the following August, lasting until July 1944. Photographed in May 1958 on the down line with a south-bound express, no. 60966 was removed from service in June 1963 and was scrapped by the end of the month.

Retford 64421

Retford first acquired railway services in 1849-50 when local services were operated out of the town, but it did not receive a connection to London until 1852. Retford's first shed was a small two-lane brick building that was completed during 1851, and at the same time a coal stage and water well were installed at the site. During 1874-75 improvements were made to the facilities; these included a new engine shed and new offices. The 1851 shed was demolished after the completion of the new shed. Gorton Works erected Class J11 0-6-0 locomotive no. 64421 in September 1907. It was originally built as part of the GCR Class 9J. The photograph was taken on 14 July 1957 and no. 64421 was withdrawn during December 1959.

Opposite below: Retford 63736

Class 04/1, 2-8-0 locomotive, no. 63736 was built by the NBLC in August 1912 to the specifications of John Robinson for the Great Central Railway. The engine was originally part of the GCR 8K Class of locomotives, which were built to handle heavy goods and in particular to transport coal to Immingham for export overseas. The 8K Class locomotives were a modification of Robinson's 8A Class, which came into service during 1902. The 8Ks had increased boiler capacity and larger diameter pistons, which improved their operating abilities over the 8As. Photographed on 14 July 1957, withdrawal for no. 63736 came after fifty-one years in service in August 1963.

Retford 64245

Doncaster Works completed the building of Class J6 0-6-0 locomotive no. 64245, by the end of October 1917. The J6 locomotives featured 5-ft-2-in.-diameter driving wheels, a boiler pressure of 170 psi and two cylinders measuring 19 by 26 inches. The locomotive weighed approximately 50 tons. Retford was the last shed for no. 64245, pictured here on 14 July 1957, before it was scrapped at Doncaster Works in March 1962.

Retford 69322

Class N5/2 0-6-2 locomotive no. 69322 entered traffic from Gorton Works during June 1899 as part of GCR Class 9F designed by Thomas Parker. A total of 131 engines were built to his specifications between 1891 and 1901, and the class had the distinction of being the first British locomotives to have the Belpaire firebox. The engine became part of Class N5 at Grouping and later N5/2 when the chimney was replaced to conform to the LNER loading gauge, which occurred in December 1934. Photographed on 14 July 1957, no. 69322 was withdrawn in June 1959. Retford shed lasted for a further six years, closing in mid-1965; the site has since become a business park.

Bawtry 60003

Upon entering service no. 4494 (later 60003), built at Doncaster Works in August 1937, was originally named *Osprey* following the practice of naming portions of locomotive classes after birds. However, by October 1942 the engine had been renamed *Andrew K. McCosh* after the LNER director and Chairman of the Locomotive Committee. This change was part of ten other changes made to the original names of the A4 Pacifics to those of men in high positions at the LNER, which occurred between 1939 and 1948. Photographed with a down line express at Bawtry during September 1958, no. 60003 was in service until December 1962.

Bawtry 60049

Gresley Class A1 Pacific 4-6-2 locomotive no. 60049 *Galtee More* is working a down line express at Bawtry in September 1958. Built at Doncaster (works no. 1604) in September 1924, *Galtee More* was rebuilt to A3 specifications in October 1945. The engine was named after the racehorse Galtee More that won eleven races between 1896 and 1897, including the 2,000 Guineas at Newmarket, the Derby and St Leger. No. 60049 was withdrawn at the end of December 1962 and scrapped at Doncaster Works during April 1963.

Above: **Bawtry 60130**

Bawtry, situated 147 miles from London and 9 miles from Doncaster Station, was also about 1 mile from Scrooby Water Troughs. The line is level at the water troughs before it rises at a gradient of 1:198 approaching Bawtry before it falls at a gradient of 1:198 between Bawtry and Rossington. The line becomes level again between Rossington and Doncaster. Peppercorn's Class A1 Pacific 4-6-2 locomotive no. 60130 *Kestrel* is on the up line at Bawtry with an express in September 1958. The engine was the first of twenty-three Peppercorn A1s to be built at Darlington Works in September 1948; the run of construction continuing until no. 60152 *Holyrood* was completed in July 1949. No. 60130 is seen on the *Harrogate Sunday Pullman* service, which started at King's Cross and went to Leeds, Bradford Exchange and Harrogate. *Kestrel*, withdrawn from service in October 1965, was scrapped by the end of the year.

Doncaster 60079
The NBLC-built Class A1 Pacific 4-6-2 locomotive no. 60079 *Bayardo* during October 1924. The engine is pictured on 12 April 1959 undergoing work at the end of 4-Bay in the Doncaster Works Crimpsall Repair Shop. Doncaster had the task of maintaining the entire class, which it received in 1930, and that work continued until late 1963 when most of the locomotives were being withdrawn.

Opposite below: **Doncaster 60063**
Class A1 Pacific 4-6-2 locomotive no. 60063 *Isinglass* is passing Doncaster South signal box whilst on the approach to Doncaster Station's down platform, which is where Bill Reed took the picture on 14 July 1957. *Isinglass* was rebuilt to A3 in April 1946 and converted to left-hand drive in November 1952. The engine is also seen with a GNR-type tender with coal rails, which is the only one it carried throughout its time in service. The Pullman car *Lydia,* seen on the left, has an interesting history. Built in 1928, the car spent time on the southern railway before the Second World War; it then became part of General Eisenhower's train when he was in Britain for the D-Day landings. *Lydia* was also part of the train that conveyed Winston Churchill to Handborough after his death. The car also spent time touring the US for many years before returning to the UK in 2000 to be preserved.

***Above:* Doncaster 60103**

Class A1 Pacific 4-6-2 locomotive no. 60103 *Flying Scotsman*, built at Doncaster in February 1923, was rebuilt with an A3 boiler in January 1947. It is seen here in front of Doncaster's mechanical coal stage on 14 July 1957. The locomotive has a streamlined non-corridor tender (with a late BR emblem), which it ran with from July 1938 until its withdrawal in January 1963. It is also carrying the 35B Grantham shed code. *Flying Scotsman* had two spells there; the first from November 1953 to June 1954 and then from August 1954 to April 1957, moving to King's Cross in both instances.

Doncaster 60116
Peppercorn's Class A1 4-6-2 Pacific locomotive no. 60116 *Hal o' the Wynd* went into traffic from Doncaster during October 1948. It is named after a man involved in a battle between two Scottish Clans in 1396, which formed part of the Sir Walter Scott novel *The Fair Maid of Perth*. The locomotive received the name in May 1951 with eight other Peppercorn A1s named after characters related to Sir Walter Scott. No. 60116 looks free from wear so it is tempting to suggest that it had emerged from Doncaster Works and was waiting to be returned to Heaton where it was based from October 1948 until September 1962. Photographed at Doncaster shed on 14 July 1957, the locomotive was withdrawn in June 1965.

Opposite below: **Doncaster 60107**
The Doncaster Carr locomotive shed, a mile south of Doncaster Station, was opened at the end of March 1876 after a great deal of work went into preparing the land for building work. Constructed of brick, the building had twelve roads and room for approximately a hundred locomotives. Class A1 4-6-2 Pacific locomotive no. 60107 *Royal Lancer* was built at Doncaster in May 1923. It is pictured on 12 May 1958, in pristine condition near Kelham Street on the north side of the shed. The engine was not on shed at Doncaster during any period of its time in service and is probably ready to return to Grantham where it was allocated at the time of this picture. *Royal Lancer* was at Grantham from September 1957 until moving to King's Cross in October 1960. Withdrawal came in September 1963.

Doncaster 60144

Doncaster Plant Works commenced operations late in 1852 after GNR locomotive repair was transferred from Boston to Doncaster. The next fifty years saw steady expansion as the work required for the railway increased. The Crimpsall Repair Shop was part of three additions to the works in the early 1900s, which also included a new tender shop and paint shop with the construction costing £294,000. The repair shop contained two small bays and four large bays and could house a hundred locomotives for repair. Peppercorn Class A1 4-6-2 Pacific locomotive no. 60144 *Kings Courier* emerged from Darlington Works in March 1949. The engine is pictured adjacent to the Doncaster Crimpsall Repair Shop, perhaps awaiting repairs or maintenance or both. The photograph dates from 14 July 1957 and no. 60144 was withdrawn in April 1963.

Doncaster 60502

Edward Thompson's Class A2/2 Pacific 4-6-2 locomotive no. 60502 *Earl Marischal* – seen here on 7 August 1960 at Doncaster Station operating a passenger service – was originally built as P2 Class 2-8-2 no. 2002 at Doncaster in October 1934. It was rebuilt to A2/2 at Doncaster in June 1944. In January 1959 the number plate was moved from its previous position above the handrails on the smokebox to where it is in this photograph, fitted to the top smokebox door hinge. This alteration was made so that the lamp irons could return to their original 1948 position. No. 60502 was the last of the class to be withdrawn in July 1961.

Doncaster 60504

Thompson's Class A2/2 Pacific 4-6-2 locomotive no. 60504 *Mons Meg*, photographed during September 1958 at Balby Bridge, Doncaster, is carrying a streamlined non-corridor tender, and it features the late British Rail emblem. As a P2, the engine operated on the Aberdeen to Edinburgh line as well as travelling to Dundee and Glasgow for periods. During its time in Scotland it was housed at Haymarket shed. After rebuilding, *Mons Meg* returned to Haymarket and stayed there for five years, operating mainly goods traffic and the occasional passenger service. In January 1950, the locomotive was transferred to Peterborough New England where it stayed until withdrawal eleven years later.

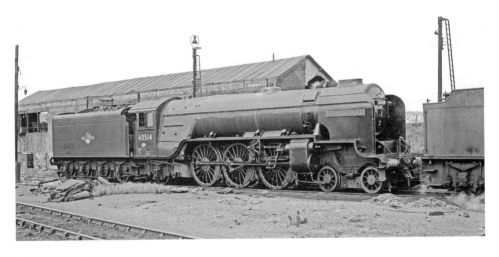

Doncaster 60514

Doncaster Carr shed originally had two coal stages at the north and south ends of the shed due to the high volume of traffic operating there. The coal stages were installed when the shed was built during 1875-76. The north coal stage was rebuilt in 1898 to improve the coaling facilities, making them less labour intensive, and to add a driver's room, which was situated underneath the stage. The north stage was retained after a new 500-ton mechanical coaler was built in 1926, while the south stage was demolished. Thompson's Class A2/3, 4-6-2 Pacific locomotive, no. 60514 *Chamossaire*, built at Doncaster in September 1946, is seen here in September 1958 in front of the old north coaling stage at Doncaster Carr shed. The engine has a plain chimney which it carried throughout its time in service; but with one exception, all the other locomotives in the A2/3 Class had their chimneys changed to the 'lipped' variety. *Chamossaire* left service in December 1962.

Doncaster 61919

Doncaster's mechanical coaling stage was installed in 1926 as part of a series of modernisations that occurred at the Carr shed. Class K3 2-6-0 locomotive no. 61919, built by Armstrong Whitworth, Newcastle-upon-Tyne, in July 1934, is pictured on 14 July 1957 in front of the mechanical coaling stage. The engine has a group standard straight-sided tender with a 4,200-gallon water capacity, space for over 7 tons of coal and is displaying an early BR emblem on the side. No. 61919, withdrawn at the end of June 1961, was scrapped at Doncaster by the end of July.

Opposite above: **Doncaster 68165**

Class Y3 0-4-0 Sentinel no. 68165 was built at Sentinel Wagon Works Ltd, Shrewsbury, during 1927. The locomotive went into service in September 1930, being employed on the Cheshire Line Committee's and Great Central Railway's goods lines at Birkenhead. The locomotive went to Doncaster Wagon Works in February 1948 to replace a Y1 Class locomotive which was scrapped. The locomotive received BR number 68165 in May 1948 and became Departmental No. 5 in March 1953. It is seen here on 22 April 1951 in front of the Weigh House at Doncaster Works with British Railways written on the side instead of the emblem. Behind is Class A3 4-6-2 locomotive no. 60072 *Sunstar*. No. 68165 was removed from service in November 1958.

Below: Doncaster 68654

Class J68 0-6-0 locomotive no. 68654, pictured at Doncaster Carr shed during September 1958, was built at Stratford Works in 1913-14 as part of a batch of ten erected during this period. The locomotive was designed by A. J. Hill, chief mechanical engineer at Stratford Works (1912-22). The locomotive was ordered by the Great Eastern Railway and was classed C72. A total of thirty engines were completed between 1912 and 1923. Although designed for shunting duties, some locomotives found their way to local passenger services. Withdrawal started in 1958 and ended in 1961. No. 68654 was part of a group of ten to leave service in 1960.

Doncaster 76105

Doncaster's New Erecting Shop was constructed in 1890-91 as the earlier facilities were becoming inadequate. The shop contained two roads running the length of the building, each accommodating five locomotives. A BR 2-6-0 Standard Class 4 locomotive no. 76105 is pictured in the New Erecting Shop at Doncaster Works on 14 July 1957. A total of 115 Standard Class 4s were built with the work being split between Horwich Works and Doncaster. The construction of the class started in 1952 and finished in 1957, with Horwich providing forty-five and Doncaster adding seventy. Designed with a view to operate freight services primarily, the locomotives also found themselves operating local passenger services all over the rail network. The Standard Class 4s were withdrawn between 1964 and 1967 with four going into preservation.

Opposite above: **York 43097**

York first opened to railway traffic in 1839 when the York & North Midland Railway opened a line to Milford where a junction was made with the Leeds and Selby railway line. The first station at York did not open until 1841 and the line was served by a temporary structure until the station was complete. The station was a built by the Y&NMR with assistance from the Great North of England Railway who were constructing a line which would link York and Gateshead. Ivatt 4F, later BR Class 4MT, 2-6-0 locomotive no. 43097 entered traffic from Darlington Works in January 1951. Photographed on 18 September 1965, the engine was in service for sixteen years before withdrawal at the end of January 1967.

Below: York 51235
Lancashire & Yorkshire Railway Class 21 0-4-0 locomotive no. 51235, pictured at York on 14 April 1957, was built in February 1906 at Horwich Works as part of sixty built in the class. At Grouping, fifty-eight became part of LMS Railway stock and the class was mainly confined to industrial areas and employed as shunting engines. By the time of the 1948 nationalisation, thirty-five of the class had been scrapped. No. 51235 joined them in October 1958; York was its final allocated shed.

York 60140

The tender attached to the Peppercorn A1s had capacity for 5,000 gallons of water and space for 9 tons of coal. Peppercorn Class A1 Pacific 4-6-2 locomotive no. 60140 *Balmoral* is in the process of taking on water at York on 13 April 1957. The engine went to York after being built and only stayed for ten months, being transferred to King's Cross in October 1949. However, this was a brief allocation lasting only eight months before the locomotive returned to York where it spent the rest of its time in operation. The Peppercorn A1s at York mainly worked passenger services to Newcastle with only one service to London operated under their steam.

York 60146

Another long-term resident, seen here running light at York on 13 August 1964, is Class A1 4-6-2 locomotive no. 60146 *Peregrine*. The engine was at 50A from June 1950 until July 1963, returning after a four-month spell at Neville Hill in October. *Peregrine* (built at Darlington in April 1949) survived for a further two years into October 1965, during which time the Peppercorns were working all kind of services in the north. As seen in this picture, diesel engines were dominating much of the traffic and steam had little time left to run.

York 60138

Before the new shed was added in 1878, York was served by five locomotive sheds. There were three roundhouses – two built in 1850 and a third added in 1863 – and two straight lane sheds, with one dating from 1841. In 1909 another shed was opened in a former York Works building close to the others known as Queen Street shed, and this group constituted York shed south. By 1937 the six sheds had become three, losing an 1850 roundhouse, Queen Street and a straight shed. The remainder were demolished in 1963. A1 Peppercorn Pacific 4-6-2 locomotive no. 60138 *Boswell*, built at Darlington in December 1948, was one of three Peppercorns to spend its entire life at York. The others were no. 60121 *Silurian*, seen here next to *Boswell*, and no. 60153 *Flamboyant*. Photographed on 13 August 1964, no. 60138 *Boswell* was withdrawn in October 1965.

York 61436

During the Second World War York became a target for the Germans and was bombed at the end of April 1942 in retaliation for the English raid on Lübeck. Sixty-five high explosive bombs fell along with fourteen incendiary bombs. York Station received a direct hit from a high explosive bomb and one incendiary. B16 no. 925 and A4 no. 4469 *Sir Ralph Wedgwood* were badly damaged in the raid and later had to be scrapped, while six sleeper cars, a lamp room, booking office, platforms one, two and three and York north shed also suffered extensive damage. B16 Class 4-6-0 locomotive no. 61436 was built at Darlington in December 1922 as no. 2365. It was previously part of the Raven-designed NER S3 Class and is seen at the south end of York Station during April 1960, with an LMS Stanier Black Five BR no. 45447 on the next platform. No. 61436 survived until September 1961.

York 60962

On the right is the old York locomotive works which never produced a large amount of engines and was mainly used for maintenance until 1905. Thereafter the buildings were used for a multitude of purposes including a gym, shed and goods storage. Gresley Class V2 2-6-2 locomotive no. 60962, built at Darlington Works in December 1942, is seen on 13 April 1957 leaving the south end of York Station with an unidentified passenger service. The engine has a single chimney and a group standard tender with the early BR emblem. No. 60962 was not fitted with a double chimney as some of the class were in the early 1960s. Withdrawal came in September 1965 after just under twenty-three years in service.

Opposite above: **York 61866**

As with other important sheds along the East Coast Main Line, the coaling facilities grew in line with the volume of traffic the shed was handling. York started with a small coal stage sited in the north shed yard. This was replaced by a larger facility in 1915, sited slightly away from the location of the original coal stage where the new number four shed had been constructed. The stage was only in use for seventeen years before a mechanical coal stage was erected, which, as with other mechanical stages, could hold up to 500 tons of coal. It was built by the Mitchell Conveyor & Transporter Company Ltd and survived until 1970, when it was demolished. Built at Darlington in October 1925, Class K3 2-6-0 locomotive no. 61866 is pictured on 13 April 1957 in York's north shed yard, with a water tower and mechanical coaling plant in the background. No. 68166 lasted until October 1961 when it was scrapped at Doncaster.

Below: York 62061

The signalling system at York was upgraded to use electric signals in 1937 and the task was carried out by the Westinghouse Brake & Signal Co. Ltd. The upgrade, covering just over 33 miles of track, took fifteen years and necessitated the closure of several signal boxes. Signalling was then handled by one signal box located in York Station. Peppercorn Class K1 2-6-0 locomotive no. 62061 was erected at the North British Locomotive Company, Glasgow, in December 1949. The engine is seen leaving the south end of York Station on 13 April 1957 with an express freight service, passing under a set of colour light signals. No. 62601 was in service for fifteen years before it was withdrawn in December 1964.

York 62758

Class D49 4-4-0 locomotive no. 62758 *The Cattistock*, built at Darlington in August 1934, is leaving the south end of York Station on 13 April 1957. The locomotive falls into the second part of the three sub-classes of D49. These engines were fitted with Lentz rotary cam operated poppet valves, while part three had and oscillating cam and part one had piston valves. The class was named after hunts or counties; *The Cattistock* was a hunt held in the Yeovil and Dorchester area. The locomotive was removed from service in December 1957. The rest of the class had been withdrawn by mid-1961 with one locomotive being preserved.

York 64706

Erected at Darlington in October 1926, Gresley's Class J39 0-6-0 locomotive no. 64706 was photographed entering the south end of York Station on 13 April 1957 while operating a local passenger service. The engine was the sixth locomotive to be constructed in a class which eventually numbered 289. The first twelve engines were different from the rest in that they were fitted with Westinghouse brakes instead of steam brakes. However, these twelve engines were brought in line with the rest when the braking system was converted to steam between 1931 and 1933, with no. 64706 being altered in May 1932. The locomotive had a further thirty years in operation before withdrawal came in March 1962.

York 65685

Class J25 0-6-0 locomotive no. 65685, erected at Darlington in December 1899, is seen on 13 April 1957 in York shed yard alongside ex-War Department Austerity 2-8-0 no. 90400. The J25 Class was designed by Wilson Worsdell for the NER, with the company classing the locomotives as P1. The engine is carrying the group standard NER tender, which had a capacity of 3,038 gallons of water and 5 tons of coal. It has four coal rails and it carries the early BR emblem. Withdrawal occurred in September 1959.

York 65894

Darlington Works built Class J27 0-6-0 locomotive no. 65894 in September 1923. Designed by Wilson Worsdell for the NER, the engines were classed P3 and a total of 115 were erected by several companies including Robert Stephenson & Co.; Beyer, Peacock and Co.; and the North British Locomotive Company. Construction of the class commenced in 1906 and lasted until 1923 when no. 65894, the last of the class, was produced. The engines began to leave service in March 1959 with the last going in September 1967. No. 65894 was bought by the North Eastern Locomotive Preservation Group in December 1967 and has been in their hands ever since, working various preservation lines and exhibitions.

York 68686

NER Class E1 0-6-0 no. 68686, designed by Worsdell and later reclassified J72, was built at Darlington in April 1899 with 4-ft-1¼-in. wheels, two cylinders measuring 17 x 24 in. and with Stephenson valve gear. The boilers had a pressure of 140 lbs per square inch, being 10 ft long and with an outside diameter of 3 ft 8 in. The tractive effort of the locomotive was 16,760 lbs, and it could hold 690 gallons of water and over 1 ton of coal. It is pictured during August 1960 at York, probably on a shunting duty. The locomotive was taken out of service in August 1961.

York 73165

British Rail Standard Class 5 4-6-0 locomotive no. 73165 was built at Doncaster in March 1957 as part of a final batch of seven engines, which would form a total of 172 locomotives in the class. The building work was split between Derby and Doncaster producing 130 and forty-two respectively. No. 73165 is seen running light leaving the south end of York Station on 13 April 1957. The locomotive was allocated to York shed when it entered service; the number of Standard Class 5s allocated there reached a total of ten by mid-1957. At York shed the locomotives worked various passenger and freight services. No. 73165 was part of a group of five that left York for Huddersfield in late 1958, where they worked similar services to Manchester and Liverpool. The engine had spells in Wakefield, Oxley and Patricroft before it left service in October 1965.

York 90079

The Austerity Class was designed by R. A. Riddles and based on Stanier's LMS 8F Class. A total of 935 were produced with 932 spending time on the continent after the D-Day landings; 733 found themselves operating under British Railways after the War, with the others finding work in Holland, Hong Kong and the USA. Austerity Class 2-8-0 locomotive no. 90079, built at NBLC's Hyde Park Works in August 1944, is pictured exiting the south end of York Station while fronting an express freight service on 13 April 1957. The locomotive remained in service until January 1964.

York D9002

BR Class 55 Deltic locomotive D9002 *The Kings Own Yorkshire Light Infantry* was built at Vulcan Foundry by English Electric. The locomotive entered service in March 1961, operating express passenger services on the East Coast Main Line. The locomotive was officially named by General Sir Roger Bower, Colonel of the KOYLI at the time, with the ceremony taking place at York on 4 April 1963. D9002 remained in service until January 1982 when it became one of the exhibits at the National Rail Museum at York. The photograph was taken on 13 August 1964.

York D2231

British Rail Class 04 0-6-0 shunting locomotive D2231, built by the Drewery Car Company Ltd, entered service during January 1956. A hundred and forty-two of the class were produced, with the building work being delegated to Vulcan Foundry and Robert Stephenson & Hawthorns. D2231 arrived at York from Darlington in mid-June 1965 and had an eighteen-month stay, leaving for Eastleigh at the beginning of 1967. The locomotive had a further two and a half years in service before it was withdrawn in July 1969. It had been scrapped at Steelbraking & Dismantling Co. Chesterfield by May 1970. D2245, seen behind D2231, has been preserved and is currently located at the Battlefield Line Railway, Leicestershire. The picture dates from 18 September 1965.

Opposite above: **York 60152**

Peppercorn Class A1 4-6-2 locomotive no. 60152 *Holyrood* was built at Darlington in July 1949 – the final locomotive to be erected in the batch of twenty-three ordered. The engine spent almost its entire time in service based in Scotland, allocated to Haymarket, Polmadie and St Margaret's, before moving to York in September 1964. It is seen here in the York shed yard looking the worse for wear, lacking its nameplate and the front number plate. Behind the locomotive is the imposing structure of the water tower. Made of reinforced concrete it could hold 100,000 gallons. It was built in 1909 and lasted until 1973, when it was no longer needed. No. 60152 survived until June 1965 and was scrapped by August.

Below: York 60124

A new station at York to replace the inadequate 1841 structure was authorised as early as 1866, but owing to a number of factors construction did not start until 1874. Further delays beset the construction, with the original contractor dropping out and being replaced by Lucas Brothers of London. The station finally opened on 25 June 1877. Class A1 Peppercorn 4-6-2 locomotive no. 60124 *Kenilworth* entered service from Doncaster Works in March 1949. The engine is seen minus its nameplates on 28 September 1965, working a passenger service at the south end of York Station. The engine was taken out of service in March 1966.

York 61049

York's north shed was built in 1878 and consisted of three roundhouses joined together. A fourth shed was added in 1915 and was attached to the north of the 1878 structure. This arrangement lasted until 1958, when numbers one and two sheds were removed and a straight building was installed; numbers three and four remained but were brought up to date. The site has since become part of the National Railway Museum. Thompson Class B1 4-6-0 locomotive no. 61049, built by the NBLC, Queens Park Works, in June 1946, went to York's north shed in mid-September 1960 and had a five-year stay before being withdrawn in November 1965. The photograph was taken on 18 September 1965.

CHAPTER THREE

Darlington to Portobello Sidings

Darlington 60076

Gresley Class A1 4-6-2 locomotive 60076 *Galopin*, built by the North British Locomotive Company in October 1924, was rebuilt to A3 in June 1941. The locomotive is seen at Darlington Station operating an express parcels train on 13 June 1962. Between 1924 and 1925, a large number of A1s were constructed and allocated along the East Coast Main Line. No. 60076, then LNER 2575, was one of ten housed at Gateshead and one of fifteen in the North East to be fitted with Westinghouse brakes (no. 60076's were removed in May 1934). The locomotive spent all its working life in the North East, the majority of the time at Gateshead, moving to Darlington in December 1948. It then regularly transferred between the two locations, completing ten moves in twelve years. A month was spent at Heaton before withdrawal came in October 1962.

Above: **Darlington 60831**

Gresley Class V2 2-6-2 locomotive no. 60831, built at Darlington Works in May 1938, was one of a number of the class that had cylinder cases altered from mid-1956. A total of seventy-one were altered due to a slight flaw in the cylinder casing. Originally the V2s had monobloc cylinders. However, if damage occurred and a patch could not be used, the whole monobloc cylinder had to be changed. Thus, it was decided to fit three individual cylinders to make repairs easier and cheaper. No. 60831's cylinders were changed in June 1957, and this is distinguishable from the position of the steam pipe to the cylinder, while the others are part of the smokebox saddle. Photographed on 18 June 1965, no. 60831 left service in December 1966; it was the last of the class to be withdrawn from the North East.

Darlington 69004

Class J72 0-6-0 locomotive no. 69004, entering traffic from Darlington in November 1949, is seen in front of Darlington Power Station on 16 June 1951. The locomotive was part of a group of twenty-eight built for British Rail between 1948 and 1951. A total of 113 were erected following the designs by Worsdell, the first of which appeared in 1898, with the BR engines being brought closer to modern standards, including new sanding apparatus and new buffers. The new engines were also approximately 1.5 ft longer and 1 ton heavier than the original engines. No. 69004 was allocated to the North East when entering service and was at Darlington for around a month in March 1962. Moving to Gateshead, the engine spent the remainder of its time in service there, being removed in September 1963.

Opposite below: **Darlington 90682**

War Department Austerity 2-8-0 locomotive no. 90682 is pictured on 18 June 1965 adjacent to the coal stage at Darlington shed. The engine was built at Vulcan Foundry which produced ten batches of locomotives in total between 1943 and 1945. No. 90682 was the last to be produced in the eighth lot, which consisted of forty-three locomotives. In total 390 Austerity Class locomotives were built at Vulcan Foundry, with the NBLC's Hyde Park and Queens Park Works contributing a further 545. No. 90682 was produced under works no. 5162 and acquired the WD no. 79219 in late 1944. It was allocated BR no. 90682 in March 1949, which it kept until withdrawn in September 1967.

Darlington 60004

Class A4 locomotive no 60004 *William Whitelaw* is pictured at Darlington awaiting entry to the works. The engine had casual light repairs carried out there between 5 July and 28 August 1965. This was the locomotive's second and final visit to Darlington Works. No. 60004's repairs and maintenance was usually carried out at Doncaster and only two other works were visited: Cowlairs in 1958 and Inverurie in 1964.

Darlington 60036

Gresley Class A3 locomotive no. 2501 (later no. 60036) *Colombo*, built at Doncaster Works during July 1934, is pictured near the turntable in Darlington shed yard on 3 October 1964. The 70-ft turntable was installed at Darlington sometime in the late 1930s to early 1940s, replacing one of 60 ft dating from 1906. No. 60036 arrived at Darlington after a six-month stay at Gateshead in December 1963. After spending almost a year at Darlington, the engine left service in November 1964.

Darlington 60045

The locomotive is seen inside Darlington straight shed, which was constructed in 1939, replacing one that had been in use since 1885. Access to the shed could be gained from both ends and it contained several roads to accommodate locomotives. Gresley Class A1 4-6-2 locomotive no. 2544 (later 60045) *Lemberg*, built at Doncaster in July 1924, became A3 in December 1927. It was only the second locomotive to undergo the alteration – after 4480 (BR 60111) *Enterprise* – with only a further four being changed before 1941. No. 60045 is pictured on 3 October 1964 with the GN tender, which it carried throughout its time in service. It was fitted with a double chimney in October 1959, and smoke deflectors were added in November 1962. The locomotive left Darlington in November 1964, after over forty years' service.

Darlington 60530

Peppercorn Class A2 4-6-2 locomotive no. 60530 *Sayajirao*, built at Doncaster Works in March 1948, is pictured on 3 October 1964 in the main erecting shop at Darlington Works. Post-1963, the maintenance of the Peppercorn A2s fell on Darlington's shoulders as Doncaster Works ceased to be involved in steam maintenance. In the foreground is the smokebox door from Thompson Class B1 no. 61330. No. 60530 spent less than two years in England in total, moving to Haymarket shed in January 1950. It spent eleven years there before departing for St Margaret's and then to Polmadie. Dundee was the final allocation in July 1964 before withdrawal came in November 1966, ending a lifespan of just eighteen years.

Opposite above: **Darlington 62007**

Built at NBLC, Queens Park Works in June 1949, Peppercorn Class K1 2-6-0 locomotive no.62007 is seen at Darlington on 3 April 1965. At the time of the photograph, the engine was allocated to 55H Neville Hill where it remained from June 1964 until October 1965. Before 1963, repairs and maintenance duties for the K1s belonged to Doncaster Works, which was visited by no. 62007 twelve times during its time in service. Four visits were made to Darlington; two between 1949 and 1950 where it underwent light repairs, and two in early 1965. The first visit stretched for two months until March 1965 when the locomotive underwent heavy intermediate repairs. The second came at the end of March when it returned for an unclassified visit. No. 62007 did not visit the works again and was scrapped in September 1967.

***Below:* Darlington 60062**
Gresley Class A1 4-6-2 locomotive no. 60062 *Minoru* is seen at Darlington on 3 October 1964 with the driver and fireman seemingly enjoying the view! The engine became A3 in July 1944 and was converted to left-hand drive in October 1952. In February 1959 it was fitted with a double chimney, and smoke deflectors were added in July 1961. The locomotive is coupled with a GN-type tender, which was reattached after it had carried a number of others. These included the new-type non-corridor from March 1937 to May 1938, and a streamlined tender until February 1946. It carried the new type again, but only for a month, before the streamlined tender was put back into operation. The GN type was reattached in February 1955 and remained until the locomotive was sent for scrap in December 1964.

Darlington D2700

British Rail Class D2 0-4-0 diesel shunting engine, built by the NBLC in July 1953, originally carried the number 11700 before receiving the pre-TOPS number D2700 around 1957/8. The locomotive was the first of a group of eight to be built between 1953 and 1956, with three coming to England and the rest remaining in Scotland. D2700 was allocated to West Hartlepool upon entering service, spending several years there working with goods and in the dock before moving to Goole in August 1960. Withdrawal came in November 1963; the locomotive is pictured at Darlington on the scrap line where it was to be cut up the following November. The photograph was taken on 3 October 1964.

Darlington 63460

Originally built for the NER and designed by Raven as Class T3, the 0-8-0 locomotive was the first of the class to emerge from Darlington Works in October 1919 as NER 901. Five locomotives were initially built with a further ten being added in 1924; they became LNER Class Q7. NER 901 acquired the BR no. 63460 in June 1951 when it was working from Tyne Dock hauling heavy mineral trains. Withdrawal came in December 1962 but the locomotive was reprieved and granted a position in the nation's collection. No. 63460 was stored at Darlington for twenty months before moving to Stratford and then Brighton. Thereafter the National Railway Museum loaned it to the North Eastern Locomotive Preservation Group, where it was fully restored and worked on the North Yorkshire Moors Railway. The locomotive is currently on display at the Darlington Railway Centre and Museum.

Gateshead 60127

Peppercorn Class A1 4-6-2 locomotive no. 60127 *Wilson Worsdell*, seen at Gateshead shed ash pit on 17 June 1965, is without its nameplates and looking the worse for wear. The locomotive was built at Doncaster in May 1949 and was named in September 1950. Upon entering service, *Wilson Worsdell* was allocated to Heaton where it worked services to Grantham, York, Peterborough, Leeds and Edinburgh. The engine moved to Tweedmouth in September 1962, working on predominantly freight services, and undertook standby duties. It arrived at Gateshead in October 1964 and continued to work various services until withdrawal came in June 1965.

Gateshead D390

The locomotive shed at Gateshead was originated by the NER around 1855. By 1875 six turntables with accommodating roads existed at Gateshead to house its engines, but by the early 1900s this had become four. An old tender shop from Gateshead Works was converted to house the larger Raven and Gresley engines in the early 1920s. No further alterations at the shed were made until the mid-1950s when two roundhouses were demolished and a large turntable was installed at the main shed for the Pacifics, as they previously used a turning triangle. The shed officially stopped housing steam locomotives in March 1965. British Rail Class 40 diesel locomotive D390, built at Vulcan Foundry, was withdrawn (as 40 190) in January 1976 and cut up in April at Crewe Works. The photograph dates from 17 June 1965.

Gateshead D9001

British Rail Class 55 Deltic locomotive D9001 *St Paddy* is seen at Gateshead shed on 17 June 1965. Even before steam left Gateshead, the shed was becoming home to the new generation of diesel locomotives; the first, a Class 40, arriving in 1960, with the first Class 55 allocated in 1961. When steam was gone, the main shed was converted into a straight shed and housed the diesel locomotives; the other buildings associated with steam were removed from the site. The site was in use for a further twenty-six years before it closed to all traffic. The site has since been cleared and is now a housing estate. D9001 was removed in January 1980, one of the first Class 55s to go, and was cut up at Doncaster Works.

Newcastle 63363

This Raven T2 Class 0-8-0 locomotive no. 1293 (later BR no. 63363), built at Darlington in June 1913 for the NER, later became part of Class Q6. The engine is photographed in the carriage sidings on 17 June 1965 with the south end of Newcastle Central Station on the left. Protruding from the station wall is the signal box that was constructed in 1959, located above platform 10, which replaced Newcastle No.1 box. The Q6s were used exclusively on mineral and goods trains because of their ability to pull extremely heavy loads. The locomotives began to leave traffic in 1963 with no.63363 being withdrawn in September 1966, while the rest had left by the end of 1967. One went into preservation.

Newcastle 65794

Newcastle and the surrounding area have a long connection with the railways extending back to the early 1800s. Local passenger services began to operate in the area during the 1830s with services going to North and South Shields and Sunderland. In 1838 the Newcastle to Carlisle line opened providing a cross-country link. The Gateshead to London line opened in 1844 and was extended to Edinburgh in 1850. Class J27 0-6-0 locomotive no. 65794 was built at Darlington Works in October 1906 as NER Class P3. Withdrawal came in June 1965.

Newcastle 03 Diesel Shunting Engine

British Rail Class 03 0-6-0 shunting locomotive D2047, built at Doncaster, is pictured at Newcastle Central Station with Turnbull's building in the background. The colour light signals in the picture replaced the semaphores in 1959. Photographed on 13 May 1965, the locomotive was withdrawn (as 03 047) in July 1979.

Newcastle 65882

Class J27 locomotive no. 65882, built at Darlington in August 1922, was withdrawn from service in September 1967. The engine was part of a group of thirty-five built between 1921 and 1923 and fitted with superheaters. These were changed to saturated boilers (which a batch of engines built between 1906-9 carried) with no.65882 changing in October 1940. When changed to saturated boilers, the length of the smokebox was reduced. This produced a noticeable difference at the front between those that were changed and those originally fitted with saturated boilers. The locomotive is also equipped with a pre-1915 tender, which is distinguishable by the frame slots being different to the later oval variety. The locomotive is seen here working a local mineral train on 19 August 1965.

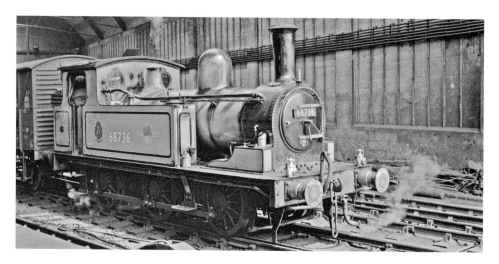

Newcastle 68736

J72 Class 06-0T locomotive no. 68736, built at Armstrong Whitworth in June 1922, is busy at Newcastle Station where it worked from 1961, acting as a carriage shunter. The engine also held this duty at York where it spent two and a half years prior to moving to Gateshead shed. The locomotive has been upgraded from the standard black, which the other locomotives in the class were painted, to a lined green livery, which had formerly been applied to NER engines. The locomotive also has the later BR emblem and the NER coat of arms painted on the left side of the tank. No. 68736 and other locomotives working similar duties received special treatment because they were on display all the time, working in the busy stations. No. 68736 held this job until October 1963 when it was sent for scrap. The picture was taken on 13 June 1962.

Newcastle 92063
Built in 1850, Newcastle Station was designed by John Dobson, principally a designer of houses and churches, for the York, Newcastle & Berwick Railway and Newcastle & Carlisle Railway. The front portico was added in 1863, and a south extension of the station was added in 1894. The station extension brought the number of platforms to fifteen, reduced to twelve in 1983. BR Standard Class 9F 2-10-0 locomotive no. 920168, built at Crewe in November 1955, has been caught reversing at the east end of Newcastle Station on 13 May 1965. No. 92068 had two spells at Tyne Dock shed, the first for only a month from November 1955, returning for a ten-year stay in May 1956; withdrawal came in November 1966.

Alnmouth 60121
Opened in July 1847, Alnmouth Station – 303 miles from London, 89 miles from Edinburgh and 35 miles from Newcastle – was originally named Bilton as this village was slightly closer to the station. It was constructed by the Newcastle & Berwick Railway Company as part of their line between Gateshead and Berwick. Peppercorn Class A1 4-6-2 locomotive no. 60121 *Silurian* is passing through Alnmouth Station in September 1964. The engine entered traffic from Doncaster in December 1948 – one of three A1s built at the works during that month – and left operations in October 1965.

Alnmouth 60982

V2 Class 2-6-2 locomotive no. 60982 was built at Darlington in June 1944 as the penultimate engine in the class. The engine is seen here at Alnmouth Station in September 1964 on an unidentified passenger service. Seen on the left is Peppercorn K1 no. 62025 on the Alnmouth to Alnwick branch line, which operated for over 116 years carrying both goods and passengers before its closure in 1968. No. 60982 was allocated to York when this picture was taken and it spent most of its life there, with brief spells at Darlington in the mid-1950s. The locomotive lasted until October 1964 when it was scrapped.

Alnmouth 62021

Peppercorn Class K1 2-6-0 locomotive no. 62021 emerged from the NBLC Works during August 1949. It is seen here in September 1964 outside Alnmouth shed, which was built in 1875 and consisted of just two roads. K1 engines were regulars working the Alnwick branch; a number were allocated in the early to mid-1960s. The shed lasted until June 1966 when it was closed and then demolished. No. 62021 arrived at Alnmouth late in 1962 and stayed until the shed closed, transferring to Sunderland for four months before a very brief stay at Tyne Dock. The engine was withdrawn in October 1966 and sent for scrap to A. Draper in Hull.

Alnmouth 62025

Peppercorn K1 2-6-0 locomotive no. 62025 was built in August 1949 – one of fourteen to appear during that month. Pictured during September 1964, the engine is working on the Alnwick branch line. Like no. 62021, the locomotive was allocated to Alnmouth at around the same time, remaining there until closure came. No. 62025 had arrived from Heaton and spent its working life in the area until it was sent to the scrap yard in April 1967. The station was named Alnmouth in 1892, changing from Bilton, but has since changed its name again; in 2003 it became Alnmouth for Alnwick. The Aln Valley Railway project is currently underway to reopen a portion of the old branch line from the station as a heritage railway.

Berwick-on-Tweed 78046

Pictured on 3 July 1961, British Rail Standard Class 2 locomotive no. 2-6-0 no. 78045 is crossing the Royal Border Bridge with a local passenger service while on the approach to the south end of Berwick-on-Tweed Station. The bridge, dating from 1850, was designed by Robert Stephenson and was the final link in the line between Edinburgh and Newcastle, spanning the River Tweed. During its time in service, no. 78045 (built at Darlington in October 1955) was a Scottish engine being housed at Hawick (at the time of the picture), Bathgate and finally St Margaret's, from where it was withdrawn in November 1966.

Burnmouth 60024

On 3 July 1961, Gresley Class A4 Pacific 4-6-2 locomotive no. 60024 *Kingfisher*, built at Doncaster in December 1936, is entering Burnmouth Station. As LNER no. 4483 the locomotive was the first A4 to be allocated to the Scottish region, arriving at Haymarket shed at the end of 1936. After a few changes, the final allocation of A4s at Haymarket was seven (60004, 60009, 60011, 60012, 60024, 60027, 60031) and these worked to Newcastle or non-stop to London King's Cross. Duties occasionally saw them operate to Glasgow, Dundee and Aberdeen. During the late 1930s, no. 60024 had spells at King's Cross and Doncaster but returned to Haymarket and did not move again until September 1963, transferring to Dalry Road, St Margaret's and Aberdeen until its withdrawal in September 1966. It was one of only two surviving A4s, the other being no. 60019 *Bittern*.

Opposite above: **Burnmouth 60891**

Burnmouth Station was opened in 1846 by the North British Railway and closed in February 1962. The Burnmouth to Eyemouth branch line was opened in 1891 and ceased operating at the same time that Burnmouth closed. Gresley V2 Class 2-6-2 locomotive no. 60891, built at Darlington in December 1939, is at Burnmouth Station with a mineral train on 3 July 1961. It has separate cylinders which it acquired in January 1957. The engine was at either Heaton or Gateshead during its time in service, and left in October 1964 to be scrapped.

Opposite below: **Burnmouth 61358**

Thompson Class B1 4-6-0 locomotive no. 61358 was erected at Darlington Works in October 1949 as part of ten locomotives built there between July and October 1949. Thompson designed the locomotives with standardisation in mind, and for the engines to be adaptable to whatever task should be presented to them. A total of 410 locomotives were built by various manufacturers. Work on the first engine commenced at Darlington in December 1942, and the last of the first ten was completed in June 1944. The NBLC completed the most locomotives, contributing 290 to the class, with Darlington adding sixty, Vulcan Foundry fifty, and Gorton Works ten. The last B1 was completed in April 1952. Photographed on 3 July 1961, no. 61358 was withdrawn in December 1963.

Burnmouth 64925

Beyer, Peacock & Co. built Gresley Class J39 0-6-0 locomotive no. 64925 in April 1937. The company constructed twenty-eight of the locomotives in the class, with Darlington contributing the rest. Building of the engines lasted from 1926 until the last one, no. 64988, left Darlington in August 1941. No. 64925 , pictured here at Burnmouth Station, left service in December 1962.

Portobello Sidings 64515

Class J35 0-6-0 locomotive no. 64515, originally no. 56, was built for the North British Railway as Class B, designed by William Reid. The engine entered service from Cowlairs Works in September 1910, one of seventy-six built in the class. A number of works were used for the construction – Cowlairs, NBLC Atlas Works, Queens Park Works and Hyde Park Works – and this lasted over a period of seven years. The locomotives were based in Scotland and designed for heavy goods duties, later finding work on mineral traffic and even some passenger services. Photographed on 18 September 1955, no. 64515 was sent for scrap in November 1961.

Portobello Sidings 64533

Class J35 0-6-0 locomotive no. 64533, erected at Cowlairs Works in March 1913, was one of the final NBR Class B locomotives to be built. After Grouping, the LNER added superheaters to the engines when new boilers were necessary. This was a long process; the last engine was not modified until July 1942. No. 64533 acquired a superheater in July 1927. The locomotive is also fitted with an NBR-style tender with the early BR emblem. It left service in January 1962. The photograph was taken on 18 September 1955.

Portobello Sidings 64590

The Class J37 0-6-0 engines were, again, designed by William Reid for the NBR as heavy goods and mineral engines. They were similar to the J35s, the main difference being that the J37s were superheated from construction and were slightly larger than the J35s. A hundred and four locomotives were built at NBLC and Cowlairs Works between 1914 and 1921. No.64590 entered traffic in December 1918 from the Atlas Works of NBLC. The locomotive fitted into the 'S' section of the class as it had a higher boiler pressure than the other engines; the 'S' locomotives boilers had a pressure of 175 psi, which was 10 psi higher than the others classed 'B'. At Grouping the pressure was raised on all engines to 180 psi. Photographed on 18 September 1955, no. 64590 was withdrawn in May 1962.

Portobello Sidings 64625

Class J37 0-6-0 locomotive no. 64625 was built at NBLC's Atlas Works in January 1921 and sub-classed 'B'. The locomotive is seen here on 18 September 1955 at Portobello Sidings with an NBR tender, capable of carrying 3,500 gallons of water and several tons of coal. Withdrawal for no. 64625 came in September 1965.

Portobello Sidings 65305

Class J36 0-6-0 locomotive no. 65305 originally belonged to NBR Class C, designed by Matthew Holmes. Construction of the class started in 1888 with no. 65305 being built in December 1898 at Cowlairs. Building ceased in 1900 when 168 were in existence. Rebuilding of the class commenced pre-Grouping (eight after), with new larger boilers and a new cab; no. 65305 was altered in December 1922. The locomotive was in service for sixty-three years, and withdrawn in February 1962.

CHAPTER FOUR

Edinburgh Dalmeny Station
to Edinburgh Waverley

Dalmeny Station 61243
An Eastfield-allocated Class B1 4-6-0 locomotive no. 61243 has crossed the Forth Rail Bridge and is approaching Dalmeny Station on 10 April 1957. The engine was one of twelve built at the NBLC's Queens Park Works in October 1947. All but one of the twelve were named after LNER directors, with no. 61243 carrying the name *Sir Harold Mitchell*. The Forth Bridge was constructed in 1890 with the GNR, MBR, and NER assembling to finance the building work. The present Dalmeny Station (initially titled Forth Bridge) opened in 1890 to serve the Forth Railway Bridge. No. 61243 was withdrawn in May 1964.

Dalry Road 42145

LMS Fairburn Class 4P 2-6-4T locomotive no. 42145 was built at Derby Works in 1950. Another LMS Fairburn 4P is seen behind the locomotive. They are passing Dalry Junction with Dalry Road Station situated on the extreme left behind the wagons. The line branches after the station with one line going to Granton and Leith and the other to Haymarket West Junction. No. 42145 is on the old Caledonian Railway Line and is heading towards Midcalder Junction, which would have taken it to either Glasgow or Carlisle. The Caledonian line and the station are no longer in existence and now form the Western Approach Road. The photograph was taken on 2 July 1961.

Dalry Road 45030

A total of 842 LMS Stanier Class 5 4-6-0 locomotives were built, starting in 1934 and not ceasing until 1951. The order for the first twenty engines went to the LMS Works at Crewe, followed by eighty produced by Vulcan Foundry. No. 45030 entered traffic from Vulcan Foundry in September 1934 and was the eleventh engine of this order to be manufactured. Further locomotives were built at these Works, augmented by fifty-four from LMS Derby Works, 120 from Horwich Works, and 327 from Armstrong Whitworth. Photographed on 2 July 1961, no. 45030 is seen at the rear of Dalry Road shed, which is where the locomotive was allocated when it was withdrawn at the end of December 1962. It was scrapped by August 1963.

Dalry Road 54478

4-4-0 locomotive no. 54478 was built in June 1920 at St Rollox Works, Glasgow, for the Caledonian Railway as Class 72 (later LMS 3P-H). The engines were designed by William Pickersgill who was Chief Mechanical Engineer for the Caledonian Railway from 1914 to 1923. A total of thirty-two were built, all becoming LMS engines, and later taken over by BR. No. 54478 was withdrawn from Dalry Road in July 1961 (the photograph was taken on 2 July 1961), and all of the class were scrapped.

Dalry Road 57565

No. 57565 was originally erected as a Caledonian Railway Class 812 (later LMS Class 3F) 0-6-0 locomotive, designed by J. F. McIntosh. Built at St Rollox Works in August 1899 as CR no. 827, the engine was one of the seventeen initially built. A further seventy-nine were erected, ending in 1909; ninety-three went to British Railways and all were withdrawn by 1964. Photographed on 2 July 1961, no. 57565 left service from Dalry Road at the end of December 1962 and has survived into preservation.

Dalry Road 57645
The shed at Dalry Road was opened in 1848 and consisted of two roads. By the turn of the century this had increased to three sheds; two with two roads and one with three. At the 1948 Grouping, the three-road shed had a fourth added, and one of the two-road sheds was demolished. Dalry Road ceased operations in October 1965. Built at St Rollox in July 1909, as a Caledonian Class 812 (later LMS Class 3F) locomotive, no. 57645 was the last one to appear. In service for fifty-three years and four months before withdrawal came, it was scrapped at Inverurie in May 1963. The photograph was taken on 2 July 1961.

Edinburgh Haymarket 60024

Gresley A4 Pacific 4-6-2 locomotive no. 60024 *Kingfisher*, running light at Haymarket shed, Edinburgh on 11 April 1957, was built at Doncaster in December 1936 and initially carried LNER no. 4483. In 1946 *Kingfisher* carried two numbers: no. 585, for two months from March until May, then no. 24 until a British Rail number was allocated in June 1948. The early BR emblem is seen on the corridor tender, a type which *Kingfisher* carried from new. The locomotive was allocated to Haymarket upon entering service and had just two short spells away; a six-month stretch at King's Cross in July 1937 and a month at Doncaster in April 1939. Twenty-four years and four months elapsed before the next move to Dalry Road then St Margaret's. *Kingfisher* was finally allocated to Aberdeen, until withdrawal in September 1966.

Edinburgh Haymarket 60031

Class A4 Pacific 4-6-2 locomotive no. 60031 *Golden Plover* was photographed on 11 April 1957 at the south side of the engine shed at Haymarket with the mechanical coaler seen to the left. Built at Doncaster in October 1937, the engine went to Haymarket, only leaving for St Rollox in February 1962. Before the Second World War, *Golden Plover* was noted as being very capable on the *Coronation* express passenger service from King's Cross to Edinburgh, which ran from 1937 to 1939. In 1939 *Golden Plover* made the trip on thirty-nine consecutive occasions accumulating well over 15,000 miles in six weeks. It also had the distinction of having the second highest number of appearances on the service, 104, beaten only by no. 4490 *Empire of India* which made the trip 125 times. *Golden Plover* left service in October 1965.

Opposite above: Edinburgh Haymarket 60041

The first Haymarket shed was built in 1842 and sited at the north end of Haymarket Station. It served the Edinburgh and Glasgow Railway Co.'s line between the two cities. A3 Pacific 4-6-2 locomotive no. 60041 *Salmon Trout*, built at Doncaster in December 1934, was the third last in the class to be erected. The picture was taken on 11 April 1957 before the engine received a double chimney and smoke deflectors, fitted in July 1959 and January 1963 respectively. *Salmon Trout* is at the east end of the locomotive shed and was the penultimate member of the class to be withdrawn in December 1965, the last being no. 60052 *Prince Palatine* in January 1966.

Below: Edinburgh Haymarket 60090
The second locomotive built as A3 at Doncaster in August 1928 was LNER no. 2744 (later BR no. 60090) *Grand Parade*. It is seen here on 11 April 1957 at Edinburgh Haymarket, on the turntable, which was located on the east side of the shed with a line to it running parallel with the south side of the shed, and also to the main line. The 70-ft-long turntable was installed in 1931-2 and built by Ransomes & Rapier. This was required because the existing facilities were proving inadequate for new locomotives such as the A3s. Also, if turning was required, the engines had to perform a circuit around the local lines. Arriving from Doncaster in July 1950, no. 60090 spent ten years at Haymarket before leaving for Dundee in November 1960. Withdrawal came in October 1963.

Edinburgh Haymarket 60098

Class A3 Pacific 4-6-2 locomotive no. 60098 *Spion Kop* entered traffic from Doncaster Works in April 1929 and is seen on 2 July 1961 at the west end of Haymarket shed, undergoing some maintenance work. With the smokebox door open we can see the components of the double chimney. LNER no. 2751 *Humorist* (later BR 60097) had a double chimney as early as 1937, although fitting them to the rest of the class only began in mid-1958. The arrangement improved steaming and made the locomotives more economical. *Spion Kop* had the arrangement attached in July 1959 but never carried smoke deflectors, which aided in the dispersal of the exhaust. Visibility problems were particularly associated with the class and the double chimney. No. 60098 left service in October 1963.

Opposite above: **Edinburgh Haymarket Station 60159**

Haymarket Station, constructed in 1842, was the finishing point of the Edinburgh to Glasgow line. It was extended to Waverley Station in 1846 but remained an important departure point. Peppercorn Class A1 Pacific 4-6-2 locomotive no. 60159 *Bonnie Dundee*, built at Doncaster in November 1949, spent fourteen years at Haymarket shed, where it was allocated from new with four others. From there, the engine was employed on the East Coast Main Line expresses and sometimes found itself on services to Aberdeen, Glasgow and Perth. *Bonnie Dundee* spent a month at St Margaret's before withdrawal in October 1963. The photograph was taken on 18 September 1955.

Below: **Edinburgh Haymarket 60516**
The general offices at Haymarket shed can be seen behind the engine and the Edinburgh to Glasgow line is almost out of view on the right. Thompson Class A2/3 Pacific 4-6-2 locomotive no. 60516 *Hycilla*, built at Doncaster Works during November 1946, is seen on 11 April 1957 using the turntable. The engine spent time at Heaton and Gateshead before going to York. Allocation of the other Thompson Class A2/3 engines, when entering service, saw four go to King's Cross and one to Haymarket (no. 60519). *Hycilla* was withdrawn in November 1962; three of the class in Scotland managed to survive into 1965.

Edinburgh Haymarket 61243

With the opening of the Forth Bridge in the early 1890s, Edinburgh saw an explosion in the amount of traffic that was using the railways around the city. This caused quite a problem and meant that the lines had to be expanded with better facilities provided. The present Haymarket engine shed was commissioned in late 1891, with building work starting the following year. It is situated just over half a mile away from the early shed at Haymarket Station. After initial problems the work was completed in 1894, providing eight through roads. Thompson Class B1 4-6-0 locomotive no.61243 *Sir Harold Mitchell* was built at the Queen's Park Works of the NBLC in October 1947. The engine is on the lines at the south of the shed, with the turntable and the general offices in the background. Photographed on 18 September 1955, no. 61243 left service in May 1964.

Edinburgh Haymarket 62427

Class D30/2 4-4-0 locomotive was built for the NBR as J Class and designed by W. Reid. Erected at Cowlairs Works in July 1914 as NBR no. 418, the engine's number changed to LNER no. 9418 in August 1924. In November 1946 this changed to no. 2427 with the BR number allocated in November 1948. The locomotive was named *Dumbiedykes* after the character in the Sir Walter Scott novel *The Heart of Midlothian*. At Grouping, the LNER split the class into two groups – D30/1 and D30/2. The first group of two had Schmidt superheaters, while the other group of twenty-five had Robinson superheaters. However, when boiler renewals were required, several members of the two groups did not receive their original-type superheater. No. 62427 was withdrawn in April 1959 and scrapped by September. The photograph was taken on 11 April 1957.

Edinburgh Haymarket 62719

Gresley Class D49/1 4-4-0 locomotive no. 62719 *Peebles-shire* was built at Darlington in May 1928 as part of the first batch of twenty engines erected there. Photographed on the turntable at Haymarket shed on 11 April 1957, the locomotive was a long-term resident at Haymarket, arriving in March 1943 and staying until November 1959. No. 62719 was originally fitted with a group standard tender with capacity of 4,200 gallons and 7.5 tons. However, during the Second World War, shortages led to twenty-eight locomotives receiving GC-type tenders and the original tenders attached to O2 and K3 Class engines. The exchanged tenders were similar in appearance with few minor differences and had 200 gallons and 1 ton lower capacity. The tender is seen here with the early BR emblem. The locomotive was withdrawn in January 1960.

Opposite above: Edinburgh Haymarket 62743

Class D49/2 4-4-0 locomotive no. 62743 *The Cleveland* entered traffic from Darlington in August 1932. The engine was one of forty D49s named after hunts and all of these fell in the second part of the class. The locomotive was also a regular fixture at Haymarket between January 1951 and May 1960 when it was withdrawn. The D49s allocated there worked a variety of services including local and express passenger services, goods workings and as pilots. No. 62743 is seen on 11 April 1957 at the west side of the shed with its original group standard tender and displaying an early BR emblem.

Opposite below: Edinburgh Haymarket 65251

The mechanical coal plant seen in the background was added in the early 1930s to replace a manual facility sited in the same position. The design for the structure came from Henry Lees & Co. and was similar to the ones seen at King's Cross, Grantham, Doncaster and York, being operated in the same way and having a capacity of 500 tons. The coaler was demolished when diesel locomotives took over the shed in around 1965. Class J36 0-6-0 locomotive no. 65251, built by Sharp, Stewart & Co. in February 1892, is seen on 2 July 1961 in the yard on the west side of the shed with an unidentified A3 receiving some attention to the left. The J36 left service in November 1963.

Edinburgh Haymarket 65243

Class J36 0-6-0 locomotive no. 65243 was erected at Neilson & Co. in December 1891 as NBR C
Class. During the First World War, twenty-five of the class saw service in France, and no. 65243 (as
NBR no. 673) was one of them. To commemorate this involvement, the engines were named after
important figures and battles of the conflict. No. 65242 received the name *Maude* after Lieutenant
General Sir Frederick Stanley Maude who was prominent in the Middle East and took Baghdad
in 1917. He also died there after contracting cholera in 1917. From Haymarket, *Maude* worked
local services, local goods and as a pilot engine. Leaving service in 1966, no. 65243 went into
preservation with the Scottish Railway Preservation Society and is currently on loan to the NRM.
The photograph was taken on 2 July 1961.

Edinburgh Haymarket Station 67615

Class V3 2-6-2T locomotive no. 67615 is running light at Haymarket Station on 18 September 1955. Since this scene was captured many changes have occurred in and around the station area. The building on the left, behind the wagons, has been demolished and replaced, first by a car park and then offices with a new dead-end platform installed in front of them, opened in 2006. The canopies over the platforms have been modernised and the only remaining feature is the station building looming over the platform. No. 67615 was built at Doncaster in June 1931 as a V1 Class engine, designed by Gresley. A total of eighty-two V1s were built, with ten being of a modified design and an increased boiler pressure. These were classified as V3. When boiler renewals were necessary, V1 engines received the V3 boiler and were subsequently reclassified. No. 67615 was modified in this way in December 1953, and operated until December 1962 when it was removed from service.

Edinburgh Haymarket 60161

Peppercorn A1 no. 60161 *North British* was the penultimate engine to be built in the class at Doncaster in December 1949. The locomotive received its name in June 1951 and was part of a group of four named after companies that were amalgamated in the 1923 Grouping to form the LNER. No. 60161 is seen on 11 April 1957 at the north east side of the shed, in front of the sand kiln. The sand kiln and new workshops on the north side of the shed were installed in the 1930s and late 1940s respectively. No. 60161 was in service for less than fourteen years when it was withdrawn in October 1963 from St Margaret's shed.

Edinburgh Haymarket 68460

A total of forty Class D (LNER Class J83) 0-6-0T locomotives were built to the designs of M. Holmes; twenty at Neilson Reid & Co. and twenty at Sharp, Stewart & Co. The locomotives were predominantly used on shunting duties but occasionally they found work on local passenger and freight services. The class could be found at most LNER sheds in Scotland. No. 68460, built at Neilson Reid & Co. in April 1901, is in Haymarket shed yard on 11 September 1955 with Donaldson's Hospital, later College, in the background. The locomotive was withdrawn in November 1958 due to the introduction of diesel shunters.

Edinburgh Haymarket 69211

Diesel locomotives began to take over Haymarket shed in mid-1959 and were allocated roads six to eight. In September 1963, the last steam locomotives left Haymarket, leaving only diesels. 1965 saw one to five roads and the surrounding building renovated so the entire building was tailored to suit the diesels; such features in the yard as the turntable and coaling plant were redundant. The site is still in use today as a maintenance depot for DMUs. Construction of NBR Class A (later Class N15/1) 0-6-2T locomotive no. 69211 was completed in December 1923 at Cowlairs Works. Photographed on 2 July 1961, the engine left service in October 1962.

Opposite above: St Margaret's 60529

Peppercorn Class A2 4-6-2 locomotive no. 60529 emerged from Doncaster in February 1948 and is seen here at St Margaret's shed on 18 September 1955. The engine was among five Peppercorn A2s to be fitted with M.L.S valve regulators, with the operating rod seen running along the top of the boiler. This apparatus was attached to the locomotive to protect the superheater elements and the cylinders. It was fitted during September 1949 – at the same time that no. 60529 was fitted with a double chimney. While the valve regulators were successful in their function, they were not attached to any other Peppercorn A2s. *Pearl Diver* was in service for only fourteen years and ten months before it was withdrawn in December 1962.

Below: **St Margaret's 68348**
Class J88 (originally Class F) 0-6-0T shunting engine no. 68348 was built at Cowlairs Works in September 1919. The locomotives were designed for the NBR by W. Reid. A total of thirty-five were erected between 1904 and 1919. St Margaret's held around nine J88s and they were employed all around Edinburgh, working various shunting duties. Photographed on 18 September 1955, no. 68348 was withdrawn in August 1958; by 1962 all the engines had gone and were replaced by diesels.

St Margaret's 69144

Built by NBLC in July 1912, the o-6-oT locomotive seen here on 18 September 1955 at St Margaret's shed, was the first of twenty to be made and originally given the NBR number 907 as part of Class A. It became LNER 9907 in January 1926 and received its BR number (69144) in August 1948. At Grouping, the LNER reclassified the locomotives to N14, N15/1 or N15/2 depending on different features. N14s had a short cab and Westinghouse brakes, N15/2s had longer cabs and the same brakes while N15/1s had a large cab but steam brakes. N15/1 no. 69144 was withdrawn in February 1960.

St Margaret's 60087

The engine shed at St Margaret's dates from the mid-1840s and was built by the NBR. It was sited to the east of Edinburgh Waverley, just over a mile away. The first shed was a roundhouse, which was later joined by a straight shed and a smaller shed. Class A3 4-6-2 locomotive no. 60087 *Blenheim*, pictured at St Margaret's on 11 May 1963, was built at Doncaster in June 1930. The locomotive is seen with a double chimney and smoke deflectors that were fitted in August 1958 and February 1962 respectively. Resident at St Margaret's for two spells, first arriving in July 1960 and staying five months, no. 60087 returned in December 1961 to remain until withdrawal in October 1963.

St Margaret's 65329

Class J36 0-6-0 locomotive, no. 65329, seen at the St Margaret's coal stage on 11 May 1963, was built at Cowlairs in January 1900 – one of six built that month. The engine is carrying the Holmes standard tender, having a water capacity of 2,500 gallons and space for 6 tons of coal. It also has metal backing plates to the coal rails and is displaying the later BR emblem. Closure of St Margaret's shed came in April 1967 and the site was subsequently cleared. No. 65329 left service in December 1963.

St Margaret's D2705 and 68095

BR Class D2/1 0-4-0, previously DY11, no. D2705 was built at NBLC in September 1955. A total of eight were constructed for shunting duties, with D2705 working at St Margaret's (from new) and Leith before it was withdrawn in August 1967. It is pictured on 11 May 1963 with Y9 Class shunting locomotive no. 68095 erected at Cowlairs in 1887. The locomotive was in service for an impressive seventy-five years before it was withdrawn in December 1962 and superseded by the type of locomotive standing behind it. No. 68095 was the only engine in the class to be preserved and it is currently owned by the Scottish Railway Preservation Society as NBR no. 42.

Edinburgh Waverley 60096

The NBR first built a station in Edinburgh in 1846 called North Bridge for services to Berwick-on-Tweed. Two other stations existed in the city to serve the services of the Edinburgh & Glasgow Railway and the Edinburgh, Leith and Newhaven Railway. The NBR absorbed these companies in the mid-1860s and concentrated services from its own station, improving the facilities at the same time. Class A3 Pacific 4-6-2 locomotive no. 60096 *Papyrus* was built at Doncaster Works in March 1929 as LNER no. 2750. In the 1946 renumbering scheme, the locomotive only received one number, 96, and carried this for twenty-three months before the BR number no. 60096 was allocated in October 1948. The photograph was taken on 3 July 1961.

Edinburgh Waverley 60096

In this picture taken on 3 July 1961, no. 60096 *Papyrus* is seen from the front end in Edinburgh Waverley Station. The locomotive has received a double chimney, which was installed in July 1958, but has yet to receive the smoke deflectors that were attached in September 1961. *Papyrus* was allocated to the Scottish region for its last thirteen years in service; to Haymarket in July 1950 and then to St Margaret's in December 1961. No. 60096 came to the end of its service during September 1963.

Edinburgh Waverley 60027

Waverley Station in its present form was built in the mid to late-1890s and covers 23 acres. The construction was mainly carried out by Cunningham, Blyth & Westland. In recent years the station has undergone refurbishment. Class A4 4-6-2 locomotive no. 60027 *Merlin* is seen at Platform 16 in Edinburgh Waverley Station on 11 April 1957. A long-term Haymarket resident, *Merlin* was a regular on the express services to London and a noted performer on *The Elizabethan* summer non-stop service, which ran from the early 1950s to the early 1960s. In 1960, *Merlin* was recorded performing the trip seventy-four times with two unbroken spells (forty-six and twenty-one respectively) during the season.

Edinburgh Waverley 64624

Semaphore signals at Waverley were replaced during the mid-1930s, making redundant four signal boxes. These were replaced by two at each end of the station with the west end box still extant. The system was upgraded again in the 1970s and is now controlled from the south of the station. Class J37 0-6-0 locomotive no. 64624 was built at NBLC's Atlas Works in January 1921 for the NBR (no. 272). Photographed on 3 July 1961, no. 64624 survived until January 1966.

Edinburgh Waverley 67617
A view at the west end of Waverley Station on 11 April 1957 with the old North British Hotel, now the Balmoral Hotel in the background. Designed by W. Hamilton Beattie, the hotel was opened in 1902 and renamed and refurbished in the 1980s. 2-6-2T locomotive no. 67167 was built at Doncaster as V1 Class in August 1931, becoming V3 in October 1957. Adjacent is no. 61354, a Thompson 4-6-0 B1 built at Darlington in September 1949. Both engines were allocated to St Margaret's and in 1959 nine V1/V3s and nineteen B1s were there. No. 67617 was withdrawn in August 1962, while no. 61354 left service in April 1967.

Edinburgh Waverley 68477
Class J83 0-6-0T locomotive no. 68477 is pictured on 11 April 1957 on the east side of Waverley Station. The engine was built by Sharp, Stewart & Co. for the NBR in May 1901. After Grouping, the class was rebuilt with new boilers; one was fitted to no. 68477 in April 1924. No. 68477 also carried a later boiler (with an altered tube configuration) built in 1950 and fitted during the decade. Behind the engine is St Andrew's House, designed by Thomas S. Tait and opened in 1939 (now the Scottish Government's headquarters). No. 68477 left service in December 1962 and was the last remaining member of the class.

CHAPTER FIVE

Gleneagles to Aberdeen

Gleneagles 60004

Class A4 Pacific 4-6-2 locomotive no. 60004 *William Whitelaw* is photographed leaving the south end of Gleneagles Station on 22 August 1965. The locomotive was built at Doncaster in November 1937 as LNER no. 4462. It originally carried the name *Great Snipe* but this was changed in July 1941. The locomotive went to King's Cross when entering service, staying for a brief period before a move to the North East, being allocated to Gateshead then Heaton. In July 1941 it went to Haymarket and stayed there for twenty-one years and eight months in two spells, with three months at Aberdeen between June and September 1962. After the A4s were displaced from the ECML, a new role was found for them on the Glasgow to Aberdeen service, which is seen in operation here. No. 60004 was based at Aberdeen from June 1963 until withdrawal in July 1966.

Gleneagles 44797
Photographed on 15 May 1964, LMS Stanier Class 5MT 4-6-0 locomotive no. 44797 is entering
the south end of Gleneagles Station. The engine was manufactured as part of Lot 187 at Horwich
Works in September 1947, serving for nineteen years before being withdrawn and scrapped.

Gleneagles 45473
Gleneagles Station was opened in March 1856 as Crieff Junction and was located on the Scottish Central Railway Line between Perth and Stirling. The line branched there to the Crieff Junction Railway joining the SCR with Crieff. The CJR was taken over by the SCR in 1865, which in turn was amalgamated into the Caledonian Railway. In April 1912 the station was renamed Gleneagles with the Crieff branch line closing in July 1964. LMS Class 5MT 4-6-0 locomotive no. 45473 is seen on 15 May 1964 at the branch line platform, which is now the station car park. The locomotive left service in November 1966 and was scrapped.

Above: **Dundee 62744**

Class D49/2 4-4-0 locomotive no. 62744 *The Holderness,* built in October 1932, was photographed on 20 September 1955 in Dundee West shed yard, close to the turntable and with Dundee West Church spire in the background. The locomotive is carrying the group standard tender with an early BR emblem. The tenders for the fifteen D49/2s built between April 1932 and September 1933 originated from Class J38 locomotives as these latter did not need as high a capacity tender or ones with water scoops. Consequently, it was decided to build lower capacity tenders for the J38s without the apparatus to pick up water and switch their tenders to the D49/2s. No. 62744 was withdrawn in December 1960.

Dundee 60937

No. 60937 was the last V2 Class 2-6-2 locomotive to be built at Doncaster Works; it is pictured here in front of the mechanical coaling plant at Dundee Tay Bridge shed. The mechanical coaler was introduced in the 1930s. No. 60937 was built in March 1942 as part of ten completed at Doncaster between June 1941 and March 1942. Twenty-five more were ordered in 1941, but this was later switched with an order for 02s placed at Darlington. The locomotive went to Dundee from new and from there worked to Aberdeen, Glasgow and Edinburgh, conveying both express goods and passengers. The engine was withdrawn from St Margaret's in December 1962.

Opposite below: Dundee 62744

The Holderness Hunt is based in East Yorkshire and takes place in the area around Driffield and Beverley. No. 62744 *The Holderness*, seen on the turntable on 20 September 1955 in Dundee West shed yard, was allocated to Leeds from new. It then spent periods at Gateshead, Hull and York before transferral to Dundee from York in March 1952. Dundee West shed was built of brick in 1885 and was a through building of eight roads. It was closed in 1951 but was used for a time to store withdrawn locomotives. By 1958 it had reopened, this time to house DMUs. The site was cleared in 1985.

Dundee 60958
Dundee Tay Bridge shed was built as a replacement for a smaller shed at Esplanade Station to the
south-west. The NBR erected the new shed in around 1891 on the site seen here, which was close
to its Tay Bridge Station and also the point where the NBR line met the Caledonian Railway. The
CR shed and station, Dundee West, were close by. Class V2 2-6-2 locomotive no. 60958 emerged
from Darlington in October 1942. The locomotive is at the east end of Tay Bridge shed and was
withdrawn from St Margaret's shed in December 1962.

Opposite above: **Dundee 44677**
LMS Stanier Black Five Class 5MT 4-6-0 locomotive no. 44677 is seen heading west on 20
September 1955. Built in April 1950 at Horwich Works with Skefco roller bearings on its driving
axles, the engine was withdrawn at the end of October 1967.

Opposite below: **Dundee 64619**
Dundee West Station (formerly titled Dundee Union Street and Dundee) saw services reduced before
closure in 1965 with trains switching to Tay Bridge Station. The building was demolished for the
construction of the Tay Road Bridge. Class J37 0-6-0 locomotive no. 64619, built at the NBLC,
Atlas Works, in December 1920, is in Dundee West yard on 20 September. Dundee West Station
tower can be seen to the right in the distance. Withdrawal for no. 64619 came in December 1963.

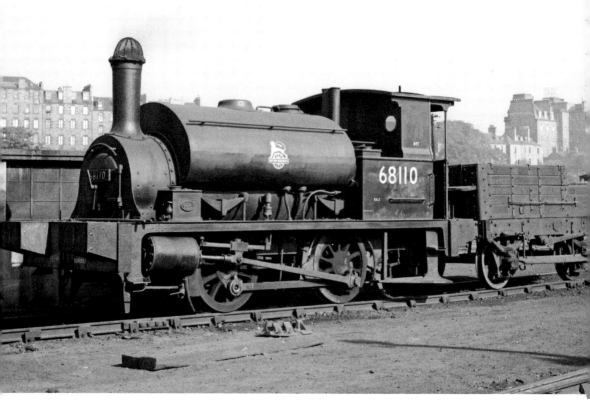

Dundee 68110

NBR Class G (later classified Y9 by the LNER), 0-4-0T locomotive, BR no. 68110 was built at Cowlairs Works in June 1891. Thirty-eight engines were erected, with three being lost before Grouping. The engine is seen on 20 September 1955 in Dundee shed yard with a wooden coal-carrying wagon, which some of the class would use as the cab bunker could not carry sufficient amounts of coal. The engine is also photographed with an enclosed cab added in May 1930; the class originally had exposed cabs. The chimney is fitted with a spark arrester as jute was present on the docks where the Y9s operated; the presence of this material posed a potent fire hazard. No. 68110 had been in operation for seventy years and two months when withdrawal came in August 1961.

Dundee 80123

British Rail Standard Class 4 2-6-4T no. 80123 is seen here leaving Dundee West Station on 10 March 1963. A hundred and fifty-five of the locomotives were built to the design of R. Riddles, with Brighton Works being the dominant builder, contributing 130. Doncaster and Derby works produced ten and fifteen respectively. Riddles based the design of the locomotives on the LMS 2-6-4T, designed by both C. Fairburn and W. Stanier. No. 80123 was built in September 1955 at Brighton Works and was in service for less than eleven years, being removed in August 1966.

Dundee 80124
Standard Class 4 2-6-4T locomotive no. 80124, photographed at Dundee Tay Bridge shed, where it was allocated from new. The engine was built in September 1955 at Brighton Works. December 1966 saw its removal from service while allocated to St Margaret's. The class differed from the LMS 2-6-4T engines in having driving wheels 1-in. smaller in diameter at 5 ft 8 in. They also had higher boiler pressure at 225 lbs, 25 lbs higher than the LMS engines, with smaller cylinders.

Opposite above: **Dundee 80090**
Riddles' Standard Class 4 2-6-4T locomotive no. 80090 was photographed on 10 May 1963 at the south-west end of Dundee Tay Bridge Station with an unidentified passenger service on 8 December 1963. The station was built by the NBR as part of the first Tay Railway Bridge project and opened to traffic on 1 June 1878. The platform, 476 yards long, was placed centrally with lines either side. Erected at Brighton Works in August 1954, no. 80090 entered service at Bury during the same month. Spells at Bangor and Birkenhead followed before the locomotive arrived at Dundee Tay Bridge in February 1960. It was withdrawn five years later during March 1965 and subsequently scrapped at Faslane.

Below: Dundee 60530

Peppercorn Class A2 Pacific 4-6-2 locomotive no. 60530 *Sayajirao* is photographed on 3 June 1966 next to Thompson Class B1 4-6-0 locomotive no. 61293 at the west end of Dundee Tay Bridge shed. The shed roof now presents a much smarter appearance than the one seen featuring Standard Class 4 2-6-4T no. 80124 as it has been replaced. No. 61293 was built at NBLC during February 1948. From Dundee, no. 61293 worked passenger trains to Glasgow and to Edinburgh, replacing the D49s on these journeys. Dundee Tay Bridge shed was closed in May 1967 with no. 60530 withdrawn from there in November 1966 and no. 61293 during August 1966.

Perth 44794

LMS Stanier Class 5MT 4-6-0 locomotive no. 44794 was built at Horwich Works in August 1947. The engine was allocated to Aberdeen Ferryhill from Polmadie shed in October 1954, staying for twelve years until moving to Perth South in December 1966. The locomotive lasted there for only four months as withdrawal came in April 1967.

Opposite above: Aberdeen 44997

LMS Stanier Class 5MT 4-6-0 no. 44997 entered service from Horwich Works in March 1947 and was withdrawn in May 1967. The locomotive was photographed on 10 May 1963 leaving the south end of Aberdeen Station (originally titled Aberdeen Joint Station) which was rebuilt in around 1915, replacing another station on the same site, constructed in 1867. The station initially served the Great North of Scotland Railway and the Caledonian Railway. Renamed Aberdeen in 1952, the station has since been refurbished and as recently as 2009 was part of the Union Square development, which saw the site undergo extensive alterations.

Opposite below: Aberdeen 60160

Peppercorn Class A1 4-6-2 locomotive no. 60160 *Auld Reekie* is at the side of the coal stage at Aberdeen Ferryhill shed on 16 June 1962. The facility was installed in around 1908 during modifications to the original shed and the site. The engine was built in December 1949 at Doncaster Works and was a Scottish engine throughout its days in service. It alternated between Haymarket and Polmadie during 1951-52 and from the latter working West Coast services before a long stay at Haymarket began. The locomotive's name is a nickname for Edinburgh. No. 60160 was withdrawn after three months at St Margaret's in December 1963, and dismantled at Darlington the following year.

Aberdeen 60005

Gresley's A4 Pacific 4-6-2 locomotive no. 4901 was named *Capercaillie* (a type of grouse) when it was built in June 1938. The engine was renamed *Charles H. Newton* in September 1942, after the Chief General Manager, and this was changed to *Sir Charles Newton* in June 1943 when he received a Knighthood. It is seen here looking in a very poor state of repair at Aberdeen Ferryhill shed on 16 April 1964, a month after it was withdrawn from service as BR no. 60005.

Opposite: **Aberdeen 60012**

Class A4 4-6-2 locomotive no. 60012 *Commonwealth of Australia*, built at Doncaster Works in June 1937, is seen leaving the south end of Aberdeen Station on 10 May 1963. To the right is the Guild Street goods yard, the site of the original Aberdeen Railway Station, which was opened in 1854 and in operation until the opening of the first joint station in 1867. From that time the site was used as a goods depot and a portion still is, while the rest of the site has been redeveloped and the station has been demolished. No. 60012 was a brief resident at Aberdeen Ferryhill shed between January and August 1964, at which time it was withdrawn. It met its end at Motherwell Machinery & Scrap early in 1965.

Aberdeen 60019

Class A4 Pacific 4-6-2 locomotive no. 60019 *Bittern*, built at Doncaster during December 1937, was a North East engine for the best part of twenty-five years (at Heaton and Gateshead). The engine made the move north of the border at the end of October 1963 with a number of other A4s. Around mid-1964 they became regulars on services between Glasgow and Aberdeen. Photographed on 2 June 1966, *Bittern* was one of the last A4s to be withdrawn in September of that year, but thankfully it was bought by Mr G. Drury and preserved. The locomotive has since changed ownership and undergone significant restoration, reverting to LNER Blue livery and no. 4464.

Aberdeen 60027

This scene at Ferryhill Junction on 16 April 1964 captures Class A4 4-6-2 locomotive no. 60027 *Merlin* south of the station and just north of Ferryhill engine shed. The junction signal box is seen in the background and was opened in 1905, closing in 1981 when the system was upgraded. *Merlin* was built at Doncaster as LNER no. 4486 in March 1937 and spent its time in service almost exclusively at Edinburgh Haymarket. While there, the engine received enamel plaques from the Captain of HMS *Merlin* in May 1946, which were mounted first on the cab and then the boiler casing, the position they are seen in here. *Merlin* worked the Glasgow to Aberdeen service and was housed at St Rollox from May 1962 to September 1964, moving to St Margaret's, from where it was withdrawn in September the following year.

Aberdeen 60919

Class V2 2-6-2 locomotive no. 60919, pictured at the Ferryhill coaling stage on 14 May 1965, was built at Darlington Works in September 1941. In October 1945 the locomotive moved from Gateshead to Ferryhill where it stayed until transferring to Dundee Tay Bridge during May 1964. From Aberdeen the locomotive hauled passenger and goods services, including fish trains. From Dundee the engine worked mineral traffic and passenger services before withdrawal in September 1966 with four other V2s. The last of the class was withdrawn in December 1966.

Aberdeen D5122

There has been a locomotive shed at the Ferryhill site since the early 1850s with improvements being made over the years as demands on the facilities grew. The shed building was a dead-end type built in 1908 with twelve roads, two being used for maintenance. It was built by the Caledonian Railway but was also used by NBR engines. Diesel locomotives were allocated to the shed in 1958 and it closed to steam in 1967. It existed as a diesel locomotive depot until closure in 1987. British Rail Class 24 diesel locomotive D5122, photographed here on 14 May 1965, was built at Derby Works in June 1960 and allocated to Inverness, operating in the Highlands. During September 1968 the engine was involved in a collision with a passenger train in which its driver and second man were killed. The locomotive was subsequently scrapped due to damage.

Aberdeen 60007
Class A4 Pacific 4-6-2 locomotive no. 60007 *Sir Nigel Gresley* is seen at the coal stage at Aberdeen Ferryhill shed on 14 May 1965. The locomotive was predominantly a King's Cross engine with a six-year spell at Grantham between April 1944 and June 1950. It was among eight A4s that were moved to New England when King's Cross closed to steam in 1963. From there, it went to St Margaret's before joining the rest of the A4 contingent at Aberdeen, seeing a further eighteen months service before withdrawal in February 1966.

Bibliography

Bolger, Paul. *BR steam Motive Power Depots*, 2009

Fawcett, Bill. *A History of North Eastern Railway Architecture Volume Two: A Mature Art*, 2003

Grafton, Peter. *Edward Thompson of the LNER*, 2007

Griffiths, Roger and John Hooper. *Great Northern Railway Engine Sheds Volume 1: Southern Area*, 2001

Griffiths, Roger and John Hooper. *Great Northern Railway Engine Sheds Volume 2: The Lincolnshire Loop, Nottinghamshire and Derbyshire*, 1996

Griffiths, Roger and John Hooper. *Great Northern Railway Engine Sheds Volume 3: Yorkshire and Lancashire*, 2000

Hoole, Ken. *Rail Centres: Newcastle*, 1986

Hoole, Ken. *North Eastern Locomotive Sheds*, 1972

Hoole, Ken. *Rail Centres: York*, 1983

Hoole, Ken. *The East Coast Main Line Since 1925*, 1977

Knox, Harry. *Haymarket Motive Power Depot, Edinburgh: A History of the depot, its Works and Locomotives 1842-2010*, 2011

Locomotives Illustrated No. 4: Peppercorn Pacifics

Locomotives Illustrated No. 9: The V2s

Locomotives Illustrated No. 10: BR Standard Pacifics

Locomotives Illustrated No. 20: Gresley Eight-Coupled Locomotives

Locomotives Illustrated No. 21: BR Standard Tank Locomotives

Locomotives Illustrated No. 25: Gresley 'A1'/'A3' Pacifics

Locomotives Illustrated No. 30: The B1 4-6-0s

Locomotives Illustrated No. 38: The LNER 'A4' Pacifics, June-August 1984

Locomotives Illustrated No. 46: The Thompson Pacifics, February-March 1986

Locomotives Illustrated No. 103: The LMS 'Royal Scot' 4-6-0s. September/October 1995.

Marshall, Peter. The Railways of Dundee, 1996

Modern Locomotives Illustrated No. 171: The Class 37's. June-July 2008

Morrison, Brian. The Power of the A4's, n.d.

Mullay, A. J. *Rail Centres: Edinburgh*, 1991

Penney, Derek. *LNER Pacifics in Colour*, 1997

Pike. S. N. MBE. *Mile by Mile on the LNER: Kings Cross Edition*, 1951

RCTS. *Locomotives of the LNER: Part 1 Preliminary Survey*, 1963

RCTS. *Locomotives of the LNER: Part 2A Tender Engines – Classes A1 to A10*, 1978

RCTS. *Locomotives of the LNER: Part 2B Tender Engines – Classes B1 to B19*, 1975

RCTS. *Locomotives of the LNER: Part 3A Tender Engines – Classes C1 to C11*, 1979

RCTS. *Locomotives of the LNER: Part 3B Tender Engines – Classes D1 to 12*, 1980

RCTS. *Locomotives of the LNER: Part 3C Tender Engines – Classes D13 to D24*, 1981

RCTS. *Locomotives of the LNER: Part 4 Tender Engines – Classes D25 to E7*, 1968

RCTS. *Locomotives of the LNER: Part 5 Tender Engines – Classes J1 to J37*, 1984

RCTS. *Locomotives of the LNER: Part 6A Tender Engines – Classes J38 to K5*, 1982

RCTS. *Locomotives of the LNER: Part 6B Tender Engines – Classes O1 to P2*, 1991

RCTS. *Locomotives of the LNER: Part 6C Tender Engines – Classes Q1 to Y10*, 1984

RCTS. *Locomotives of the LNER: Part 7 Tank Engines – Classes A5 to H2*, 1991

RCTS. *Locomotives of the LNER: Part 8A Tank Engines – Classes J50 to J70*, 1970

RCTS. *Locomotives of the LNER: Part 8B Tank Engines – Class J71 to J94*, 1971

RCTS. *Locomotives of the LNER: Part 9A Tank Engines – Classes L1 to N19*, 1977

RCTS. *Locomotives of the LNER: Part 9B Tank Engines – Classes Q1 to Z5*, 1977

RCTS. *Locomotives of the LNER: Part 10A Departmental Stock, Locomotive Sheds, Boiler and Tender Numbering*, 1991

RCTS. *British Railways Standard Steam Locomotives Volume 1: Background to Standardisation and the Pacific Classes*, 1994

RCTS. *British Railways Standard Steam Locomotives Volume 2: The 4-6-0 and 2-6-0 Classes*, 2003

Rogers, Colonel H. C. B. *Thompson and Peppercorn Locomotive Engineers*, 1979

The Railways of Aberdeen: 150 Years of History, 2000

Whiteley, J. S. and G. W. Morrison. *The Power of the A1's, A2's and A3's*, 1982

Yeadon, W. B. *Yeadon's Register of LNER Locomotives Volume Two: Gresley A4 and W1 Classes*, 2001

Yeadon, W. B. *Yeadon's Register of LNER Locomotives Volume Three: Raven, Thompson and Peppercorn Pacifics*, 2001

Yeadon, W. B. *Yeadon's Register of LNER Locomotives Volume Nine: Gresley 8-Coupled Engines Classes O1, O2, P1, P2 and U1*, 1995